INFINITE
LIFE

INFINITE
LIFE

A revolutionary story of eggs, evolution and life on earth

JULES HOWARD

Elliott&Thompson

First published 2024 by
Elliott and Thompson Limited
2 John Street
London WC1N 2ES
www.eandtbooks.com

ISBN (hardback): 978-1-78396-777-3
ISBN (trade paperback): 978-1-78396-827-5

9 8 7 6 5 4 3 2 1

A catalogue record for this book is available from
the British Library.

Typesetting: Marie Doherty
Printed by CPI Group (UK) Ltd, Croydon, CR0 4YY

'*Like a shower of stars the worlds whirl, borne along by the winds of heaven, and are carried down through immensity; suns, earths, satellites, comets, shooting stars, humanities, cradles, graves, atoms of the infinite, seconds of eternity, perpetually transform beings and things.*'

– Camille Flammarion (1842–1925)

CONTENTS

A MOST MARVELLOUS IMPERFECT
boniferous Period, 358, could say, were

PROLOGUE

Under a winking strip light in the corner of my middle infant classroom, a fish tank had been placed upon the dingy, cork-covered worktop. I can see this new addition now, in my mind's eye, from across the room. We lined up like normal that day, then planted our bottoms on the floor. As usual, the register was taken and the teacher told us about our day, yet at no point did my eyes leave the thing behind the glass: the great globule; this gelatinous dropping. This was my first encounter with frogspawn and I had never seen anything like it in my life. This mass of eggs was so strange that if you had told me an extraterrestrial visitor had broken into the class in the night and dropped this germ into our midst, I might have believed you. In the weeks that followed, I would take it all in: the dots; the twitching embryos; these assembling proto-organisms. The gathering of tiny tadpoles on the surface of the slime blob; their first lap of the tank; their second and third. The fog of my eager commentary condensed upon the glass each breaktime. Each day, my tacky fingerprints peppered the glass surface; each night, someone's job must have been to wipe them off. My quaint, rundown primary school never repeated this activity again while I

ix

was there. Probably, the teachers cursed the cleaning that having a tank of frogspawn required, as well as the smell that belched from it. But I am extremely grateful to whoever it was who said back then, in the cigarette fog of the dingy staffroom, that it might be something worth trying. I still think about those eggs today.

When you think of an egg, what do you see in your mind's eye? A chicken egg, hard-boiled? A mermaid's purse, the egg of a shark or ray, entangled in seaweed thrown onto shore? The eggs of head-lice, perhaps, being scraped off a nit comb? Perhaps you see a human egg cell, prepared on a microscope slide and telegraphed onto a TV screen in a laboratory? Or the majestic marble-blue eggs of the blackbird? Each egg is unique, and that is one of the finest things about them. Each egg on Earth has its own charisma, allure and evo-lutionary backstory, easily (I have learned) as diverse and interesting as the animals that hatch out of them. Every egg there has ever been is an emblem of survival; a product whittled, chiselled and crafted by the unthinking forces of natural selection for the purpose of passing genetic lineages forwards in time – days, weeks, months, sometimes years. Eggs have an evolutionary depth to them that animal-lovers don't consider enough. And so, *Infinite Life* is a biography, of sorts; a true history of the egg – the most unifying, resilient life structure that Earth has ever cooked up.

In the chapters of this biography, we journey through the Cambrian explosion, when animal life surged into the lineages we recognise today, when eggs were first nursed and cradled; we chart the egg's magnificent assault on the land, first through the ancestors of spiders and scorpions, then insects, then fish that first walked the shores and, later, the forests. Our journey takes in Triassic ponds, brimming with mating amphibians; the rise of maggots and other insect larvae; the marsupials thriving in newly evolved pouches and the rise of the most diminutive egg of all – the mammal egg as you and I know it. A single cell, the width of a hair, from which every human alive has passed. There is, naturally, a special place in our story for the eggs that encapsulated themselves in a layer of crystalline calcium and became the shelled eggs of some dinosaurs and, later, birds – a model example of how natural selection can modify egg shape, structure, colour and form and how beauty can manifest itself upon nature without design or deity. But the evolution of the bird egg is just one story among hundreds of others that feature here.

To write this evolutionary biography has, at times, been challenging. It took time to shift my perspective and view animals as bit-part players in the story of life, for once. I have been writing books about animals for fifteen years, yet this is the first time I have found myself having to look past them, at this new, earlier, evolutionary frontier. In time, as I worked through life's chapters, writing from this new perspective became clearer. In fact, by the end of *Infinite Life*, I began to see animals as little more than vehicles to make more eggs, which of course, in an evolutionary way, they are. However, there are times in this book when it may seem as if eggs have desires, wants or needs: that eggs *wanted* to move from the sea to the land; that eggs *sought safety* in the mammal uterus or *hid themselves* in crystalline bird eggshells. Nevertheless, eggs – devoid of a brain and incapable of an instructive thought – can clearly do very little else than simply be an egg. Eggs are not capable of knowing their journey. I give them agency at particular moments only to better tell an engaging story.

Another challenge, which likewise required careful navigation, was on which animal eggs to focus. After all, there are many millions of animal species alive today and millions more that lived in the past. I decided to choose as diverse a spread of eggs as possible – from insects to mammals, from sea urchins to sharks – 'zooming' in and out of their lives to best describe the ways that the egg was revolutionising itself. My goal was to explore how evolutionary changes in the egg affected, shaped even, animals and their ecosystems through time: from before the Cambrian Period, before animals as we know them today existed, through the Silurian Period and Devonian Period, when coastlines shifted and climates see-sawed; into the Carboniferous, where bony land animals made advancements across continents; then to the Age of Reptiles – the Triassic, Jurassic, Cretaceous – and into the modern day, where mammals are now having their turn at the top. Eggs were there, the whole way, evolving in a host of incredible directions.

Some evolutionary 'flashpoints' in egg history include the evolution of free-swimming sperm and egg in the midst of the pre-Ediacaran Period; the evolution of the 'soma' ('body'), the seat of animal mortality; the evolution of internal fertilisation and the implications this had for the sex lives of land animals. And, more recently, the evolution of the placenta, whose scar, once a fracture point between mother and offspring, sits a few inches above your lap.

A further challenge in the writing was the fact that, at many points in Earth's history, very little is known about what some eggs looked like. The reason for this is that the fossil record is, by and large, biased towards hard things – bones, teeth, claws, shells. The things that fossilise well. Eggshell is, in many cases, in most species, too soft to fossilise, and so most prehistoric eggs have been lost to the sands and muds (and death-consuming micro-organisms) of time. To fill in any gaps in our knowledge about eggs in evolutionary history, I have looked to closely related modern-day animal species for clues and hints or, where DNA analysis allows, I have investigated some of the secrets found deep in animal genomes. That makes *Infinite Life*, in general terms, more of a gentle journey (or a 'thought experiment' even) through evolutionary history than an exhaustive and exacting account of every egg that there ever was. Ultimately, the question I hope this book answers is: *can we re-frame the story of animal evolution through the lens of the egg?* It has been a great privilege to try.

As a child and as an adult now, the joy of eggs is that they sit between the boundary lands of life and death. They represent potential. That's what I see in frogspawn each year to this day – an exciting, dizzying, extreme form of potential. And so, for me, the egg will always be what I saw in that school tank all those years ago. Something to smear a nose up against; to question; to express wonderment at; to find new perspectives of life in.

Eggs are, I now realise, the mechanism through which animal lineages are propelled forwards through time, like threads woven into a giant tapestry. Eggs are the stitches. Eggs hold together life generations and lineages; they bind the animal narratives known to us today, allowing us to marvel both at the big picture and the intricate needlework of life. Told from the perspective of the egg, there are surely things we can learn about what it is to be alive.

Jules Howard

1

DUST FROM DUST

*Hadean Eon, 4,540 million years
ago, to the end of the Cryogenian
Period, 635 million years ago*

'But if (and oh what a big if) we could conceive in some warm little
pond with all sort of ammonia and phosphoric salts, — light, heat,
electricity present, that a protein compound was chemically formed,
ready to undergo still more complex changes . . .'

— Charles Darwin, in a letter to Joseph Hooker (1871)

Among the billions of stars taking shape in our galaxy long
ago, the yellow dwarf sun that would come to light our
days would not have immediately stood out as anything
unique. In the long view, it was just part of a constellation, nothing more. Another pinprick of light, born from a spinning disc of
super-heated matter, like almost everything else. Out of the planets
that formed around this disc, one single rocky sphere near the middle of the pack started to develop differently to any other we know
about so far in the universe. On Earth, the sun energised oceans and
continents. Our planet soaked up its radiation and things started to
stir. In time, life would evolve here. And with it, in the life lineage
that would become animals, came the egg.

How far back does the egg go in Earth's history? Right to the
start? Not quite. At first, 4.5 billion years ago, this place was barely
a planet at all. It was more like a condensing cloud forming from
debris – ice, rock, dust – that circled our fledgling star. It was an

aggregation of matter, one of many aggregations orbiting our Sun. Earth would have looked like a misshapen blob in those earliest millennia, pulled out of shape by the gravity of the objects caught in its grip. But as this sphere grew, its gravitational pull increased and, slowly, it became rockier.

There was nothing gentle about those early Earth days. Not a single rotation passed without the bombardment of new rocks and ice and new showerings of space dust. Some chunks of rock were as big as countries are today, some as big as continents and some, rarely, were even bigger still. Although the explosions caused by their impact were violent and tumultuous, these intemperate collisions, without any thought or forward planning, delivered the precursors for life – the elements, especially metals, that would become the building blocks of proteins and the structural units of cells and their membranes. These gifts from the solar system are in you now. Many comets in this early period delivered ice, which vapourised upon impact. In our paper-thin atmosphere, water molecules grouped together again, coalesced and, for the first time, formed beads at the mercy of gravity. Clouds formed. Rains fell. There was something akin to weather.

Cauterised through millions of years of bombardment, the Earth's surface became a hellish melting pot, but there was no stopping the maelstrom of chemical interactions now. Elements shuffled against one another to form minerals which would go on to become extra ingredients in life's cauldron. Unstable isotopes of uranium sank deep into the Earth, stoking a fire that still rages beneath our feet today, and the potential of carbon, an atom which can unite with other elements in multiple arrangements, was being explored through a trillion or more chanceful, random interactions. Some carbon arrangements held for moments. Others, in the form of the oldest diamonds, were forged 100 kilometres below the Earth's crust at temperatures of up to 1,200 Celsius and they are still locked in their arrangements as you read these words.

In that first few hundred million years, Earth's atmosphere was gossamer thin. Yet, as its crust hardened, gases began to belch out from the chaos churning kilometres beneath. These gases burst upwards through turbulent volcanic eruptions, through cracks and fumaroles. In the sea, chemicals boiled from vast seams that must have looked like open wounds in the ocean floor. Some of these produced ocean vents, where water heated from deep underground thrust itself upwards into the cold ocean waters. Here, where hot met cold, minerals in the water deposited themselves into tubes that loomed upwards like monolithic organ pipes in a vast cathedral.

The Earth is one character in this, our setting of the stage for the egg. But there is another character too. It is Luna – our moon.

Luna also has history. It begins as a cosmic congealment of dust caused by an explosive collision unrivalled in our history, when Earth was struck by a planet-sized object named Theia. Some 6,500 kilometres wide, Theia is likely to have had its own molten core and, perhaps, water. After engaging itself with Earth's orbit, Theia glanced across the face of the planet, at close to a 45-degree angle. This happened very early on in our history, perhaps only a few hundred million years after the Earth began to form. The collision could so easily have been a near miss. The cloud of debris that formed after the connection of these two celestial objects was so great that, once pulled back into shape by the workings of gravity, it became our Moon. The collision affected the Earth forever, knocking its rotation off its axis, causing it to spin at an angle. From this point onwards, Earth would spin 23.5 degrees off kilter. This is the first of a number of flashpoint moments for the egg, for this wonky spin meant that our planet ended up with an exaggerated seasonal change – spring, summer, autumn, winter. Seasons bring with them chaotic periods of ebb and flow; feast and famine; life and death. And in time, eggs would evolve in ways that would navigate animals, and their genetic material, through hard times like these. But we are getting ahead of ourselves. For billions of years back then, life was mostly about relentless reproduction – single-celled organisms dividing and dividing, throwing variations to the anarchy of the oceans; a sea of losers, speckled with accidental winners, dodging the certainty of uncertain environmental change.

Evidence for these early micro-organisms abounds in numerous forms, many going back 3 billion years or more. This evidence comes from biogenic graphite, loosely described as the fossilised remnants of early cells left as chemical imprints on graphite and carbonate, dug out of rock faces by geologists. Littered within these samples are carbon arrangements derived from organic reactions common to life today. Evidence also comes from garnet crystals, formed in the Earth's earliest pressure-cooker environments, that contain carbon molecules preserved in a chemical death grip with other atoms seen commonly in biological systems today. These biological arrangements are composed of oxygen, nitrogen and even phosphorus, a common component of cell membranes and internal cellular components, including DNA. Indeed, rock impressions from the Pilbara region of Western Australia contain pyrite-bearing sandstones that have within them micro-fossils of tube-like cells once thought to oxidise sulphur using a primitive form of photosynthesis. Life was, to turn a phrase, finding a way 3.7 billion years ago.

But there were no eggs or anything much like them at that time.

There is evidence from 200 million years later than this – an eyeblink in evolutionary terms – of cement-like structures (made by micro-organisms) known as stromatolites. Stromatolite colonies were so large back then that they could have been stood upon like giant slimy toadstools, protruding from coastal pools and lagoons. Like weary soldiers, survivors of evolutionary wars untold, these ancient organisms remain in a handful of tidal lagoons today (most famously at Shark Bay in Australia) where salinity remains so high that grazing animals such as sea snails stand no chance of sustaining a slimy foothold. These glistening, bulbous stromatolites are formed from centuries of microbial growth, each generation building upon the last in an orgy of photosynthesis. But, again, eggs remain absent from this diorama of early life.

By 1.9 billion years ago, cyanobacteria were thriving. Gorging on sunlight and producing energy for growth in the presence of carbon dioxide, these micro-organisms flourished in surface waters. Great clouds of oxygen, their waste product, began to cause our atmosphere to change. Minerals in rocks exposed to the atmosphere began to oxidise, stripping oxygen atoms from the atmosphere to create flaky red deposits rich in iron ores that are still found today. The Earth's surface, literally, started to rust. Oxygen levels rose. The colour of our planet changed. Within another evolutionary blink of the eye, these photosynthetic organisms had painted parts of the Earth's surface in greens, reds and blues, giving life to water in slimy mats upon the shores. Through cyanobacteria, for the first time, the planet Earth had been endowed with signature colours visible from light years away.

These vast populations of cyanobacteria, many trillions strong, started to evolve. Through accumulations of mutational copying errors, individuals and their populations became slightly different from one another and it was upon this variation that the cogs of natural selection bit. As the sun's radiation poured across the Earth's surface, the rate of these copying errors waxed and waned. With each incoming deluge of photons, the broad-brush forces of natural selection set to work, removing from the gene-pool those most ill-equipped for survival. Seasons must have come and gone like impeccably timed plagues back then. And in the shadow of their retreat, the ones that were left – the toughest, the hardiest – continued their evolution. There was no higher aim or purpose to any of this. No 'desire' to make something new or better. It was just that those individual micro-organisms best at not dying began to fill up the seas. The world was, in other words, beginning to host a growing accumulation of successful mistakes.

In one group of single-celled organisms, the cell's structure evolved and became more complicated. This sort of cell had within it a range of smaller sac-like structures, one of which, the nucleus, would provide the housing for DNA. These were the 'eukaryotes' – the ancestors of plants, fungi and animals, organisms that (sometimes but not always) freely swap DNA between them through a phenomenon we call sex.

This wasn't sex as we know it today. Instead, back then, for eukaryotic cells, sex was a far more primitive activity, probably involving large free-living cells engulfing smaller free-living cells and assimilating some or all of their genes in the process. Once it took off, this simple form of sex happened probably quite predictably at certain times or in certain seasons. At other times of year, these individuals reproduced in an 'asexual' manner too, the classical style, with single cells splitting in two to produce identical, daughter cells.

And so, although there was a simple form of sex 1,000 million years ago, there was no true egg, defined as an organic vessel grown by an individual to carry offspring to term. The egg is a very 'animal' endeavour. It was far too soon for that.

That's not to say that some of these cells did not become very egg-like, of course. Between 1 and 2 billion years ago, a very resilient form of vessel started to evolve. A resting stage (or 'resting cyst') evolved in cyanobacteria. This was an armoured sleepsuit within which organisms could see out hard times. In life, before becoming these resting stages, these single-celled organisms were likely to have lived like modern-day plankton, floating in surface waters, using energy from the sun's rays to grow.

Their resting stages are visible in the fossil record, clear as day. The tiny fossils are identified and studied by the most diligent, yet underappreciated, of all fossil scientists – the palynologists.

Palynologists (from the Greek verb *paluno*, meaning 'dust') obtain these tiny egg-like fossils from fine-grained shales and silt-stones by bathing them in hydrochloric or hydrofluoric acids for days on end. The acids strip away the minerals from the rock face and create a gooey residue of what was once organic matter. Within this residue, sieved and sieved again with ever finer mesh sizes, are the hard remains of resting cysts, the walls of their cellular ramparts toughened with a layer of complex molecules that resemble those found in modern pollen grains. A single 25-gram sample of rock can contain hundreds of these peculiar microfossils. A single sample could occupy an entire PhD.

Fossilised cysts really do have a lasting charm. When first released from their rocky matrix and spilled onto microscope slides for closer examination, palynologists rarely know how many there will be or what types of cyst they might see. Magnified by forty or one hundred times, the microscope slide becomes like a two-dimensional continent. In fact, so huge is the slide in comparison with the fossilised cyst that, when a potential suspect is spotted, its location on the slab of glass is noted with a special grid reference. Each microscope slide is like a historical map, with each grain of dust its ornate treasure.

Under a light microscope, resting cysts resemble seeds. Their spherical surfaces have upon them patterns of ridges and furrows which look as if they have been carved in drying cement. Most contain within them a hollow internal chamber in which a single unicellular organism once lived. The coverings of some specimens appear almost mesh-like. Or corrugated, like roofing. There are pores visible in many. Some cysts come complete with rows of tiny portholes. It is through these pores that, many millions of years ago, the environment influenced the micro-organism, now lost.

Using a scanning electron microscope to provide far better magnification, the intricacy of the cyst surface becomes stunningly complex. There are ripples and folds of protein-laden sheets that overlap with one another like leaves on a cabbage. Some specimens are covered in distinct patterns of dimples or, as they evolved and diversified, lumps and spikes. Not all examples are spherical. Far from it. Some are cube-like; some are adorned with wisps and ribbon-like structures. In life, scientists argue, such arrangements may have helped these organisms float in the surface waters of the long-ago world. Now, like beads in a broken necklace, they litter the late-Proterozoic rock layers.

Many scientists subscribe to the view that the cysts are ancestors of today's dinoflagellates – unicellular planktonic life forms that bear a resemblance to baroque shields, powered by a pair of asymmetrical whip-like ribbons that provide the engine for movement.

In the modern day, dinoflagellates are best known for causing 'red tides', where populations increase exponentially and, within the space of days and weeks, turn surface waters into a ghoulish red mist. At this point, when dinoflagellate populations peak and resources deplete, starvation ensues across the population and the inevitable die-off begins. Not all dinoflagellates perish in these moments, however; many become resting cysts. In this hibernation chamber, the dinoflagellate sinks to the sea floor, ready to re-activate when the good times come back; when Earth, at the behest of the seasons, shall provide. Many scientists suspect that it was the same for the ancient 'cysts' described from fossils.

For a billion years, up until about 650 million years ago, the humble cyst is an enduring theme of the fossil record. A hardened wall, mostly impermeable to environmental chaos, which maintains an internal environment that can be controlled. A private space, tightly maintained to ensure the survival of molecular threads, conglomerations of enzymes and oozing cytoplasmic subspaces.

The resting cyst was a device that propelled genetic material forwards in time. The product of a sun, a moon, a changing planet riddled with environmental uncertainty. In the presence of death, the egg – as a concept – was forged. Yet, only in the biological kingdom known as Animalia would it become so refined, so natural, so crucial a part of the life and life story of Earth organisms.

And so, in the millions of years that followed the evolution of the resting cyst, a new sphere-shaped structure began to fall upon sea floors across our world. This vessel contained cells which divided, over and over, which coalesced into the shape of an organism, a unique genetic individual that could, by activating muscles and a primitive brain, escape its casing and live, truly, a new life.

The egg, as you and I know it today, was coming.

2

THE GARDEN
OF MORTALITY

*Ediacaran Period, 635 million years
ago to 538.8 million years ago*

'Omnis cellula e cellula.'

('Each cell is of the cell.')

— Rudolf Virchow (1821–1902)

Imagine an art gallery where great works are not categorised by century or by geography and not ordered by chronology. A gallery in which the art styles of thousands of years are caked, daubed and plastered upon one another; where famous works can be added to and tinkered with by others that follow. For more than 60,000 human generations, that art has resided upon the rocks of Namibia's Aar plateau, one of the most important geological strata in the world for understanding pre-animal life. Traditionally this incredible landscape was home to the San people, one of the first indigenous hunter–gatherer cultures in southern Africa. The San came to these rocks because, when impacted with a pointed rock, a white dot is produced upon the dark grey limestone background. These dots, when collected into shapes and patterns – of snakes, zebras, antelope, elephants – are the expression of human minds past. They are still added to today, creating a miasmic mood board of the human experience over time. And interspersed between and under these artworks are fossils. It is upon these rock faces that the so-called Ediacaran Garden has been exposed, where details of the organisms that lived in the geological age before animals as we know them today existed. When palaeontologists study this layer of rocks, they see a jumbled mess of creation – a biological puzzle, still being solved, featuring strange and non-sensical biological beings whose fossilised lives are yet to be fully understood.

Some Ediacaran organisms, like *Charnia*, resembled living feathers, collecting nutrients (we assume) using globular lobe-like structures. Others, we think, were worm-like and dug branching burrows through the sand. Some were large, like *Dickinsonia*, whose fossil imprints show it to have been the size of a pillowcase, engorged and close to bursting, perhaps, with some unseen, viscous fluid. Did it have eyes? A stomach? Gills? No one really knows. Among these curious Ediacaran organisms are fleeting glimpses in the fossil record of one of the earliest true eggs. Seeing or studying these eggs requires a steady hand, endless patience and an extraordinary eye for detail.

For Ediacaran eggs to avail themselves and be studied properly, the rock samples first need to be partially ground up and sliced into thin sections, some one twentieth of a millimetre thick, then mounted on slides for examination under 200 per cent magnification. To see them at their best, polarised light is often used, which lights up the crystal elements of each egg, helping scientists to gauge their orientation. The polarised light gives a cosmic quality to the images that the scientists produce. Each tiny sac looks like a distant supernova or a far-off galaxy, bordered on all sides by a ring formed from some forgotten explosion. As with NASA images, there is an element of confusion and chaos here too, mostly because the Ediacaran eggs are far from intact. Each is battered and fragmented, so much so that some members of the palaeontological community still deny that they are eggs at all.

Most notably, in many of the 120 types described, there is the suggestion that, at the time of their death, when they were fossilised and frozen in time, the cells in these eggs were forming a recognisable embryo. In fact, some fossils seem to show cells at a stage in embryonic development known as gastrulation, when the embryo transforms from being a hollow ball of cells (known as the blastula) into a three-dimensional cup-shaped structure (the gastrula) from which development continues. Some of these specimens appear to show the tiny fossil eggs in familiar states of division: four-celled structures, for instance, and eight cells, sixteen cells and so on. These, surely, were true eggs – embryonic cells, contained within a protective envelope, starting to develop into new life.

But what produced these eggs? And are they comparable to those of any modern-day species?

Some fossil specimens, particularly those from China's Doushantuo Formation, show Ediacaran eggs with a defined single cell layer (the ectoderm) and a clear inner (endodermal) layer, folded and divided into finger-like protrusions that resemble the early stages of the free-swimming larvae of jellyfish and other representatives of their taxonomic group, the Cnidaria. Many suspect this is where the eggs came from. Jellyfish and their kin, the cnidarians, are, quite probably, the first true, most profligate, egg-layers.

With no blood and a limited nervous system, jellyfish have an obvious 'before-animal' feel about them. They seem to fit right in among the Ediacaran organisms. They digest things with a simple pouch-like bag, with one single opening; their internal and external worlds are divided by a body wall, just a mere cell or two in thickness. Compared to animals as we recognise them today, jellyfish make no sense at all. But, in the context of the Ediacaran, with its miasma of blobs, slime and semi-symmetrical splurge, the cnidarians fit right in. Today, almost 15,000 cnidarian species are known. The group includes corals, sea anemones, sea pens, the box jellies, hydrozoans (which include all freshwater cnidarians as well as the free-floating Portuguese man o' war) and the 'true' jellyfish known the world over. The latter include in their ranks some that measure barely a few millimetres in diameter and others with bell diameters of more than 2 metres, and tentacles that can extend up to 30 metres in length.

Cnidarian eggs are small, sometimes the size of a full stop, sometimes less than the width of a single human hair. Under a microscope, their eggs appear as tiny, gelatinous spheres, occasionally with a slightly granular, sometimes speckled, veneer. But eggs are a tiny part of the life cycle of this primitive animal group. For great chunks of time, sometimes for years, these organisms reproduce asexually, without the need for egg or sperm. This may well have been, for many Ediacaran organisms, for most of the time, the way that things were done back then.

Many cnidarians live in colonies in which every individual is a clone of their neighbour. These colonies are extended by individuals 'budding' themselves – releasing into the water small amounts of body tissues that fix themselves to nearby rocks and grow into completely new, genetically identical, individuals. Here they are known as 'polyps' – a cnidarian life stage where the organism consists of a body, attached to the rock face via a stalk, with a single upward-pointing opening surrounded by multiple tentacles. In colonies of cnidarians, hundreds of polyps can become gathered together, producing a tangled carpet of silky tentacles that appear to lick and flick at each passing plankton-filled current. As the colony grows and conditions in the water change, often seasonally because of food availability, the young cnidarians ready themselves for a new stage: they produce free-swimming forms (that look like true jellyfish) known as medusae. In many (but not all) species, the tiny medusae drift off to make a life in the open sea where they become, essentially, free-swimming sex machines – sporadic purveyors of sperm and eggs to ensure each lineage continues.

It is in sunlit waters that the medusae produce eggs and sperm and where sexual reproduction takes place, externally, in clouds of activity unseeable to the human eye. Once fertilisation occurs, the egg, its cells undergoing their earliest divisions, drifts downward onto the sea floor where it becomes a new polyp, which begins to develop and 'bud' into 'daughters' that go on to become a new population of asexually reproducing clones. Thus, these organisms go through waves of both sexual and asexual reproduction, as many Ediacaran animals may have done. They produced true eggs, in other words, fertilised by a new character: sperm.

The significance of this cannot be understated. Although the distant ancestors of animals (shared with plants and fungi long ago) were able to transfer their genetic material, now, here in the Ediacaran, was a gang of (in many cases) multicellular interlopers actively growing and expelling cells that could mix with others in the waters to grow entirely new genetic individuals. The egg had evolved, but it was not to be a solitary character. In this era, sperm evolved to become its accomplice.

It is hard to say what those early sperm may have resembled. Their small size is probably ancestral – maybe once, tail-less, they actively bumbled into eggs, 'urging' the larger egg cell to engulf them entirely so that DNA could be incorporated and a simple form of fertilisation could occur. Regardless, from the earliest days of the animal egg, sperm were there. Proof of this can be seen clearly in the genomes of all modern-day animals. The gene that humans rely on to activate sperm production (known as the Boule gene) is found in animals as distantly related as insects, crustaceans, starfish and snails. Jellyfish (and other cnidarians) also possess the Boule gene, suggesting that, together, we inherited this gene from a common ancestor from which we collectively evolved, more than 600 million years ago. Thus, the instruction manual for making sperm and eggs is, almost certainly, an early (there or thereabouts) Ediacaran innovation.

The universal propensity for animals to make sperm in the same manner, using the same genetic tools, is not readily disputed by scientists. What is disputed, on the other hand, is why sex became more and more important to so many animal lineages at this time. Because, when you think about it, sex makes no sense at all. That an animal would spend so much effort, engaging a mate, producing eggs, producing sperm, to attempt to ensure just 50 per cent of its genes make it into the next generation . . .? You would expect natural selection to correct that quickly. One would assume that in such a world of sex, where individuals are working tirelessly to pass on just half of their genes, that cheats would prosper – that those individual organisms that focus solely on producing eggs with 100 per cent of the individual's genes, without any need for sperm, would flood gene pools and flourish. But this is almost universally not what we see in animal lineages. At some point around the Ediacaran, some of the earliest animal groups (among them, the distant relatives of insects, spiders, fish and many more) appeared to invest more heavily in sex as a means of reproduction, pulling away from the budding (asexual) strategy championed by jellyfish and others. So why? Why sex?

Science writer Carl Zimmer describes the trio of jostling theories for the origins of sex as the 'good', the 'bad' and the 'ugly'. Like playing cards being shuffled, sex opens up the possibility of 'good' hands cropping up in populations, which natural selection favours. This, over millions of years, leads to the generation of successful, adaptable organisms, far superior to the asexual organisms that are left behind in their wake. So that's the good, but what of the bad? In asexual organisms, a naturally occurring random mutation can be lethal, which is, in no uncertain terms, a very bad thing for genetic lineages. If an individual acquires a mutation before asexually reproducing – a corrupted cell molecule working comparatively inefficiently, say – then every daughter cell will have the same mutation. 'Bad' genes are inherited, passed down, ad infinitum. There is no way to flush these bad 'cards' from the population and so they persist. Worse than that, through additional mutations, more bad genes can join them. Compare that to sex, where individuals share cards, swapping the best ones around, while natural selection flushes the worst hands from the deck round after round.

The 'ugly' in this trio of ideas to explain why sexual reproduction evolved at all refers to parasitic invaders – often in the form of fungal attackers, viruses and bacteria. If a parasite stumbles upon a method to, say, break through the cell wall of its host, that same trick will work upon the entire population, since they are all, genetically, the same. This means that a single beneficial mutation in an individual parasite can see the descendants of the lucky parasite immobilise a population of asexually reproducing organisms, sometimes even eradicating them totally. By mixing genes up, sex is constantly changing the passwords in populations, shuffling the codes, refreshing and updating the security software to keep the invaders at bay.

Good. Bad. Ugly. We have no way to know which it was that led to the favouring of more and more sex in Earth's first garden, 600 million years ago. It may well have been a combination of all three.

What we can see, up close and in startling detail, are the fossilised eggs ejected into the water and fertilised in this pre-animal period. The battered, fragmented cosmic sacs cemented into their mudstone beds. The semblance of cells – two, four, eight, sixteen – of lost, un-extrapolated early animals tinkering, for the first time, with sperm and eggs.

Where once organisms were using egg-like structures as opportune containers to send genes forwards in time, in the Ediacaran Period eggs became vehicles for genetic mixing. In this period, eggs became cradles for new life, new growth and, crucially, for new combinations of genes.

Back then, most of these Ediacaran organisms may have moved between stages of asexual reproduction and sexual reproduction, as jellyfish do today. But in the period that followed, the strains of early life that leaned into sex more fully, often dedicating their entire life cycle to its purpose, became the ones that would diversify most spectacularly. They would become the crustaceans, the molluscs, the earliest fish, the worms and others. And so these new eggs were, in a non-literal sense, a fuse or a spark; they became the backlight for the evolutionary explosion to follow. The Ediacaran Period was the sexual awakening of animals. The Cambrian Period was its new dawn. A new era of reproduction, which continues to this day, began. Evolutionarily, sperm and egg, together, won out.

3

THE EARLY WOMB

*Cambrian Period, 538.8 million years
ago to 485.4 million years ago*

'*Ab ovo usque ad mala.*'

('From the egg to the apples.')

— Horace's *Satire* 1.3

Looking upward through the water from our position on the floor of a shimmering Cambrian lagoon, the red sun flickers between moments of definition and distortion. Each rippling wave appears to usher the burning star towards the western horizon; further and further still, until the sky darkens. As it dips out of sight, a huge star-lit moon, some 20,000 kilometres closer than it is today, begins to bleach the wave tips. Its ghostly blue light stabs its way downwards into the deeper depths, energising the animals lying there. Waiting for this night, for this moment, to begin.

At first, from a distance, it looks as if a ghostly veil has been draped over the sea floor. It is silk-like, see-through and it shimmers. Currents in the water gently move the strange cloud, causing it to slowly stretch and twist as if caught in a breeze. Moving closer, we realise that this illusion is caused by the shimmering of a million tiny granules in the water. It is a mist of animal eggs, caught in the moon's luminous gaze. And surrounding these eggs is a swarm of single-celled activity an order of magnitude smaller. It is a frenzy of motile sperm, their whipping tails swimming in a sea within a sea.

27

These eggs look passive, as if strewn randomly across the water column, but they are not. They are active participants in sex. Although we cannot see it, in the jelly layer of each egg a siren song is being produced. Its notes are chemicals, which blend into a rich chorus that penetrates its surroundings. As they drift through the water these chemical notes slip through the cell membrane of the nearby sperm cells, causing pores upon their surface to open up. Calcium ions flood into the cell now. This influx of ions influences the chemical pathways that govern sperm motility, causing the movements of the sperm's tail to become asymmetrical, thereby bending the sperm's swimming path towards the egg. Captured by the egg's song, the sperm homes in.

This is one of a million primitive miracles occurring in the waters on this moonlit Cambrian night, at the dawn of what scientists often like to call the 'Cambrian Explosion' – a 13-million-year period in which a host of recognisable animal groups evolved, including worms, crustaceans, brachiopods (shellfishes unrelated to most modern day species), molluscs, starfish and primitive fish. This was a time when sea levels rose, flooding low-lying parts of rocky continents, allowing opportunistic animals of the ocean more territory into which to expand and more voluminous sunlit waters for photosynthetic algae to thrive in. This period of warmth saw glaciers retreat and there were no longer frozen regions at the Earth's poles. Many continents, at this point devoid of obvious plants, were surrounded by tropical shallow waters in which grew reefs made mostly from cyanobacteria and colonies of simple, sponge-like organisms. As these lifeless continents moved, occasionally they collided, generating intense volcanic activity that lifted, folded, crumpled great expanses of land into vast mountain ranges that would one day erode back into the sea, without ever being scaled by a living thing. Because life, in the Cambrian Period, was mostly a water-based phenomenon.

More spawning takes place now. New eggs are flushed into the sea from animals that resemble sea urchins resting below on a lime-green reef. Each egg, from a distance, is little more than a dot. Closer, much closer, each egg is a shiny orb, coated in shimmering fur. The strange visual effect upon the surface of each egg comes from its covering of molecular strands – the 'zona pellucida'. Each 'hair' upon this furry coat is made from a single branching ribbon-like protein (a glycoprotein) awaiting connection with a surface molecule found only on the front end of the sperm of its species. This glycoprotein is faithful to only one combination of molecules – the fate of a genetic lineage rests upon it.

In this Cambrian scene, a sea urchin sperm makes contact with an egg. The sperm, firmly entwined in the egg's tentacle-like surface, wriggles frantically as if to bury deeper into its surface. As it does so, its head begins to fall apart. Its burrowing surface appears to crumble. Bit by bit, the sperm begins to digest itself, courtesy of enzymes released in their first moment of connection. During these frantic struggles, new enzymes wash against the surface of the egg, shattering the ribbon-like proteins as they shift and twirl around it. It may look, in these energetic struggles, as if the sperm is opening a path through the egg's protective membrane, but look more closely and one sees that, again, the egg is an active agent in this union. Thousands of tiny finger-like extensions ('microvilli') wrap around the sperm, engulfing the denaturing head, holding it in place. As these fingers begin to be reabsorbed, they pull the sperm inward. Although popular culture has it that the sperm somehow 'penetrates' the egg with the frantic wriggling of its tail, the truth is that the tail offers almost no propellant force by this stage. After all, a single sperm, its tail motoring with all the energy it can muster, can be tethered in place by something as tiny as a single molecular bond, stretching out from the surface of the egg. Sperm is intrepid, perhaps, but weak and fragile. It is the egg's embrace that determines its fate. And so, both genetic vehicles in this complex dance have character and agency.

As the connection between egg and sperm deepens, a tsunami of chemical changes washes outwards across the egg's surface, again caused by a chain reaction driven by the sudden movement of calcium ions. This sudden change in ion concentration within the egg causes sac-like organelles known as cortical granules to burst, releasing enzymes across its surface which cause the egg's furry covering to rapidly fall apart, obstructing the approach of any other sperm in the area. Once these cortical granules have ruptured, they can never be restored.

As the egg surface shuts its doors to intruders, the sperm's genetic information begins to move through the egg cell to complete its objective of genetic transfer. Within a matter of seconds, almost like a protective airbag being inflated, the egg's cell surface begins to lift off from its surroundings, forming a structure known as the fertilisation envelope. Upon this inflated surface, new molecules bind with one another to form a hardened matrix that helps the egg resist being shattered by its surroundings. It is within this protective structure that the newly forming embryo will develop – deriving order from the chaos of fragmented glycoprotein chains, ruptured granules and dead sperm.

In days and weeks, the cells of this embryo will divide and divide again and, as they do so, the cells inside it will begin to specialise. Some will become digestive tissues, others sensory; some will be involved in respiration; some for armour, some for spines and some for venom. And there, sectioned off and secure, will be the cells that will one day become egg follicles – where animal eggs are made and, during ovulation, released. In this way, the 'cell lines' of the developing embryo become separated into two forms: one group of cells specialises and becomes body tissue cells (or 'soma' cells); the other group does not specialise in the same way, instead becoming cells that produce eggs or sperm – the 'germ' cells. Genetically, never the twain shall meet again.

This sectioning-off of eggs early in the formation of the embryo became widespread in animals of the early Cambrian or just before. In fact, all of the major animal groups that live on land today organise their embryos this way. It seems like a small detail, this quirk of cellular organisation, but it was far more than that. So important is this way of doing things that our day-to-day decisions, our animal mortality, our very way of life, are influenced by it.

An explanation is required.

To understand the significance of this cellular arrangement, we must travel back to the animals of this Cambrian world. Picture them as two rival groups: Montagues and Capulets, if you will. There are those that section off their egg-producing cells early in the embryo stage (the ancestors of crustaceans, insects, molluscs and all vertebrate, back-boned, animals) and there are those that don't do this, instead producing eggs later in life by re-jigging their existing body cells to turn them into egg- or sperm-producing organs. Most notably, this group includes the jellyfish and other cnidarians, whose roots (as we have seen in previous chapters) go back to the Ediacaran. Jellyfish and their close cousins manage this cellular 're-wiring' trick through stem cells, which are special cells able to produce lines of daughter cells of a range of types and functions — neurons, stinging cells, eggs, etc. Sometimes, more than half of the cells in their bodies are stem cells. Should an organism like this be cut in half, it can regrow a completely new body, which can go on to make eggs and sperm. Most animals today, including humans, are very much not this kind of organism. From the day we are born, we lack the stem cells to organise the regrowth of legs or arms or fingers or toes, for instance, should we lose them. Unlike some cnidarians, we cannot 'bud off' a few body parts and watch them regrow into egg-producing clones, more's the pity.

Like sex, this arrangement doesn't immediately make sense. You might think, for instance, that if we did have more stem cells in our bodies today that we would have better survival odds. That the genes of budding humans would proliferate in gene pools. And yet, perhaps gratefully, that is decidedly not what we see. Even insects, which are very diverse in their reproductive strategies, have not unlocked a 'budding' mode of life. There must have been something good about 'going mortal' in those pre-Cambrian primordial seas — so what was it?

A trio of hypotheses attempts to explain why cell lines become so wedded to separating eggs from body cells so early in development. The first hypothesis is that separating out the cells that will later produce eggs and sperm (the 'germ cells') early in development 'unlocks' a faster rate of evolution in other body parts, causing organisms to adapt and evolve more quickly, often evolving into new species when they do so. Supporting this hypothesis is the fact that frogs, which section off their egg-producing cells very early in development, appear to evolve and become new species ('speciate') far more readily than salamanders and newts, which do not. (This difference in evolutionary approach leads to an interesting flipside for salamanders and newts: by wielding stem cells later in life, these organisms can regrow limbs and eyes and even parts of the brain, should they be lost.)

The second, more traditional, hypothesis is that separating out of germ cell lines and body cell lines early in the development of the embryo reduces the number of cell divisions required to make eggs and sperm, reducing the likelihood of mutations finding their way into future generations. The argument goes that by reducing the number of cell divisions in germ lines, the risk of copying errors is significantly reduced.

More recently, a third hypothesis has entered the mix: that mitochondria, tiny organisms free-living within animal cells, are the driving force behind the early sectioning-off of the germ line. By way of a reminder, mitochondria are cellular hitchhikers, distant ancestors of the free-living cyanobacteria, that, in the presence of oxygen, give energy to cellular process. Many animal cells contain thousands of mitochondria in each cell, so vital is their work. But mitochondria cannot 'jump ship' between organisms, whether closely or distantly related. Instead, mitochondria journey from individual to individual only through the medium of animal eggs.

This means that every mitochondrion in your body is directly descended from the mitochondria that lived, like a rag-tag collection of islanders, in the egg that made you. Those mitochondria were put into that egg, long before you were born, deep inside the body of your birth mother.

This third hypothesis has it that limiting the cell divisions of germ cells is a good way for complex animals to limit damage to mitochondrial, rather than (or as well as) animal, DNA. This mitochondrial smoking gun, outlined by Nick Lane in the brilliant *The Vital Question*, is given weight by the observation that animals with low mitochondrial mutation rates, such as sponges, tend not to invest in germ lines early. Bluntly, they don't need to because they are under less intense evolutionary pressure to do so. It is worth noting that, in these 'primitive' species, evolution appears to run at a pace ten or twenty times slower than other animals, such as crustaceans and molluscs, that were evolving in the early stages of the Cambrian Period.

There really are big day-to-day repercussions to this strange approach to how our embryos develop. You wear a seat belt or put on a cycling helmet partly because of an evolutionary deal struck long ago in this ancient part of Earth's history. Until medical science improves, we are likely to remain in this mortal shell.

Could this strange sectioning off of eggs so early in embryonic development have been part of the reason for the kick-starting of the Cambrian explosion in so many animal lineages? It's possible. Many palaeontologists agree that the burst of new animal forms first seen during this period is likely to have had a variety of fuses. Some scientists argue that the evolution of new animal species was powered by the rise in atmospheric oxygen occurring at the time. Others argue that Cambrian animals were enriched by excessive calcium carbonate – a vital ingredient of armour, spines, mandibles, shells and bones – that was washing into the oceans at the time from the rocky continents. There are scientists who have argued that the evolution of predators upped the evolutionary ante back then, driving the evolution of survival techniques and, by extension, new animal species. Or that the evolution of eyes, as we know them today, unlocked new evolutionary avenues for natural selection to exploit. It is highly likely to be a combination of these factors. But the division of germ (egg and sperm) and soma (body), and the impact this had on the body's approach to making eggs, is surely another of the fires being lit in those early days.

The germ line, then, is the first of two big egg innovations that occurred in the Cambrian. The second egg innovation is easier to see, given that it is preserved, sometimes in pristine detail, upon the surface of sediment-rich Cambrian rocks. Some animals, fossils tell us, were moving away from the strategy of pumping their eggs blindly into open water as their ancestors did. Instead, they began to invest in caring for their eggs by attaching them to objects in their environment.

The details of these early animals are recorded upon rocks collected from high up on Mount Stephen, Canada, from the world-famous Burgess Shale, where the significance of the Cambrian explosion was first realised. Upon these 515-million-year-old slabs play out forgotten scenes of fallen Cambrian empires, consumed in sediment so thick and fast flowing that bacteria and other organisms back then had no opportunity to settle into the task of decomposition. Today, these shale slabs are mostly stored in the US National Museum in Washington, DC.

As with Ediacaran layers of sediment, Cambrian rock layers contain remains of eggs and embryos that palynologists (dust-studiers) recognise. Many are found littered across the ancient sea floor like spent betting slips or blemished lottery scratchcards. And, in a way, that's what they are. It is clear when looking at the markings and ethereal etchings on the Burgess Shale, that in the Cambrian Period there begins some toying with new techniques to tip survival odds in the egg's favour. Just as animal lineages seemed to explode in their diversity during this period, so too did their eggs.

Some of these spectacularly well-preserved fossils show that eggs were beginning to become attached to things, protected, apparently, from rushing currents or bruising storms. By the middle of the Cambrian Period, some trilobites – the arachnid-like super-group that would dominate the world's oceans for hundreds of millions of years – were settling upon this technique. These crustaceans resembled modern-day woodlice, with spiny protrusions on their upper surface. Some crawled on the sea floor, some burrowed, some swam through the surface waters. Although it is thought that trilobites were largely external fertilisers, like sea urchins, some groups were clearly starting to find safer locations for their eggs.

Fossils from the Kaili Formation in southern China show trilobite eggs laid in clutches of seven, arranged in a hexagon shape on the substrate, like tightly packed snooker balls. Each pearl-like egg is about half a centimetre in diameter; each cluster no bigger than a small coin. There is nothing random about these clusters. The 2-3-2 arrangement of their eggs has been shown through simulations to direct water currents favourably across their surface, halving the likelihood of the eggs being wrenched from the rock face during storms and high tides.

Another significant change to the egg during the Cambrian Period was its placement in an entirely new surrounding – in the protective, cradling embrace of the 'marsupium' – a brood pouch.

The word 'pouch' when applied to animals conjures up an image of a kangaroo or wallaby – a young joey peeking out from a pocket on the front of the body, where it spends its early life protected. But kangaroos, wallabies and other marsupial mammals are latecomers to this evolutionary party trick, for pouches are an evolved morphological adaptation that has been around for many millions of years and which evolved independently in many species. In the modern day, there are frogs (in the genus *Gastrotheca*) with marsupia made from a folded cup of skin, extended upwards from the base of the frog's back; there are fishes with marsupia on their belly, including seahorses, razorfish, pipefish and seadragons. Among crustaceans, even in animals as widespread as woodlice, marsupia are very common indeed. The reason that this adaptation so commonly occurs in animals is simple: if natural selection can endow animals with eyes and ears it surely has no trouble enlarging a sheet of skin or body armour to make a protective shield for the protection of eggs and embryos.

The earliest marsupium – the first protective womb – was therefore a Cambrian innovation, developed most comprehensively among the woodlouse's distant crustacean ancestors. *Chuandianella ovata*, recovered from the Maotianshan Shales of China, is one example of a marsupium-endowed crustacean. This shrimp-like

Cambrian character had a long, fluttering tail and a body encased in what looks like the shell of a broad bean. *Chuandianella* was an active swimmer that fed using feathery, filtering structures protruding from underneath its armoured carapace. The crustacean attached one hundred or so eggs, each barely half a millimetre in diameter, to the inside surface of its body armour and therefore its body armour doubled up as a mobile safehouse for eggs.

Chuandianella was not alone. *Waptia fieldensis*, a similar shrimp-like crustacean with a more rotund body shape that reached almost 8 centimetres in length, also hit upon this strategy. This arthropod, one of the first Cambrian fossils discovered in the Burgess Shale in 1909, also reared its eggs internally. Its eggs, 2 millimetres in diameter, are larger than those of *Chuandianella*, yet the clutch sizes are small – just two dozen or so. Again, these eggs were attached to the inner surface of the armoured carapace. Then there is *Kunmingella douvillei*, another early crustacean, a blur of whirring legs atop a hinged protective covering. *Kunmingella* carried upon some of its pairs of legs noticeable egg sacs. In each case, the fossilised eggs are spherical and pea-like, made shiny in their preparation. Again, its eggs differed from *Chuandianella* and *Waptia* in size, number and arrangement.

This is an important point, for Cambrian eggs began to routinely diverge between species: they were big and small; numerous and few. The establishment of the germ line was the first big Cambrian egg innovation, the second was that eggs were now attaching to things. But there was a third innovation: an 'experimentation' in number and variety of eggs.

In the Cambrian Period, animals began, evolutionarily speaking, to weigh up the 'many or the few', the 'little or the large' of eggs, sizing up the strategies that gave the best returns and the most comprehensive genetic legacies. On one side was what we often think of as 'primitive' creatures: corals, jellyfish, the earliest trilobites and sea urchins. These animals pumped hundreds or thousands of small eggs into the water, 'knowing' that at least one or two might survive. They are what ecologists sometimes call r-strategists, the 'r' referring to 'rate' in their equations. On the other side are the organisms which were being pulled to another extreme. Instead of investing in lots of betting slips or lottery scratchcards (to continue the metaphor) these organisms produced fewer eggs but invested more resources in each one, increasing their survival chances. Although the eggs were less numerous, each one was larger, the embryo often afforded more energy to grow by way of extra energy reserves. Today, in modern animals, we can see that bets like these, though risky, have a good chance of paying out in certain environments. Humans, dolphins, whales and elephants are notable proponents of this strategy. We are K-strategists, the 'K' referring to *Kapazitätsgrenze* ('capacity limit'). *Kunmingella*, *Waptia* and *Chuandianella* were among the earliest Cambrian groups to (metaphorically) peek at the survival odds on offer and change their bet accordingly – these animals were, quite probably, some of the world's earliest animal K-strategists.

What brought about this evolution in brood care? What variables were changing in the Cambrian oceans that led to pouches becoming a winning strategy? Back then, it could have been down to the evolution of new predators, with sensitive eyes and monstrous mouthparts. It may be that eggs kept in pouches were better maintained with life-giving oxygen by the whirring limbs situated, anatomically, close by; or that the increasingly complex food webs generating themselves in the Late Cambrian created more stable environments where risky bets (in the form of parental investment) were guaranteed to succeed more frequently. Whatever the driver, eggs in pouches became, in a very real way, the safe, secure option.

This is not to give the impression that all of the animals of the Cambrian were investing in protecting their eggs in body armour or attaching them to rocks. Many animals of the time remained broadcast spawners. These included early molluscs, including snail-like creatures and representatives of the shellfish, for instance, sand-dwelling worms and, of course, many of the cnidarians, including jellyfish. And let us not forget the sea urchins with whom this chapter began. When studied today, using high-powered microscopes, it is impossible not to be impressed at the level of specialisation that occurs upon each surface. These eggs were, and remain, survival machines – for me to refer to them as 'primitive' earlier is deeply unfair.

Broadcast spawning is far from a simple evolutionary choice. Free-floating eggs are uniquely adapted for survival, picking up a host of further adaptations along the way that can be seen clearly today in species that spawn directly into water. Just because this condition is ancestral does not make it un-evolved.

From the first moments they are released, in the Cambrian or in the modern day, broadcasted eggs face a range of threats which they are well practised at combatting. Looming high among these threats is the sun, which brings the threat of solar radiation in two specific wavelength bands: UV-B and UV-A. These two subversive forms of radiation, when rained down upon the sea, have the capacity to raise hell in embryos for the simple reason that they damage thymine molecules, one of the four 'bases' from which DNA is made. The resulting miscreant thymine molecule that UV radiation can create is almost always quickly removed by other cell proteins that break them down and rebuild them with replacement molecules. The problem is, in cells that divide again and again and again, *almost* always is not good enough. Eventually, harmful mutations will accrue, killing or severely impacting the function of the cell and each of its subsequent daughters. Because of the radiation threat generated by a burning star 150 million kilometres away, eggs evolved specialist molecules to perform the role of sun shading.

The most important, and evolutionarily ancestral, molecules for this purpose are the mycosporines and mycosporine-like amino acids (MAAs) which are found in organisms as diverse as cyanobacteria, fungi, seaweeds, lichens and animals, particularly those found in freshwater and marine environments. These molecules work in much the same way as inorganic and organic sunscreens and sun lotions: their role is to absorb photons of UV light of specific wavelengths, crumble apart and dissipate the radiation harmlessly as heat. Every floating egg cell of billions comes packaged up with its own supply of MAAs, which the adult organism unthinkingly divides out into each and every egg like a caring note left in a child's lunchbox. Each dose must last long enough for the offspring to develop and escape the solar storm hitting the ocean's surface.

UV radiation brings with it other problems for developing cells. These wavelengths have a nasty habit of making oxygen molecules less stable – promoting the formation of oxygen atoms with unpaired electrons (known as free radicals) which, as they move around cells, strip electrons from proteins and other important cell molecules.

To disempower free radicals, embryos are able to employ specific proteins released and maintained early in life, including glutathione and ovothiol, whose job it is to control their spread.

There are other threats that Cambrian eggs must have evolved to navigate in their earliest moments, including rampaging bacteria, for whom eggs offer an easy feeding substrate. The first and most widely relied upon adaptation to keep bacteria out is the evolution of a physical boundary. In many species that boundary is the fertilisation envelope – 10 per cent of the total protein in each sea urchin egg is dedicated to this purpose, so strong is the selection pressure to form a tough boundary to deter invasive agents.

Eggs also have chemical weapons that take the form of 'defensins', a large group of tightly folded proteins, typically with a high positive charge capable of disrupting other complex molecules. These were probably also there in the earliest eggs, produced in their millions, during the Cambrian Period. Egg cells abound with defensins, which can cause a range of protective actions. Some defensins disrupt the cell membranes of their invaders, causing them to flood with water and eventually burst. Others stop bacteria from being able to regenerate damaged cell walls or act to neutralise toxic products secreted by invaders. As the cells of the embryo develop, the surrounding egg membrane becomes primed with these molecules, which stand guard like watchful soldiers on the ramparts of a castle. Fungal infections have inspired in eggs other adaptations aimed at protecting the developing embryo. Sometimes, cultures of 'friendly' bacteria are encouraged onto the surface of the egg as a 'blood sacrifice' for approaching fungi. These cultures are introduced to each egg upon their release from the female's body.

Should the cell wall be violated during these encounters, egg cells are also capable of sealing up wounds. If an object pierces through its defences, sacs in the cytoplasm (activated by a change in calcium concentration) fuse with one another to make a hard plug. As the cytoplasm gushes out of the cell wound, the plug is carried to the scene of the cell wall violation where it gently and effortlessly seals up the site of intrusion. This is why scientists can tease a syringe into a cell and pull it out without any lasting damage. These sacs are especially numerous in the eggs of marine invertebrates, underlining the threats that such eggs faced in their history, adapting as they did to violent marine environments, including those that existed more than 500 million years ago.

And so, through these numerous adaptations, the Cambrian Period was as much an explosion for eggs as it was animals. This period saw eggs become cradled and protected in primordial wombs. In others, eggs became evermore polished, floating survival vessels. It was during this time that the egg perfected its siren song for sperm; that its defensive walls were cemented; when deals were done about size and number; when evolutionary bets were taken and, perhaps most profoundly, where the roots of our mortal bodies were secured.

Moonlit nights in complex Cambrian seas saw the egg take new forms, drawn down novel evolutionary avenues. That there would be a subsequent explosion onto the land, upon which nutritious plants and fungus were evolving, was entirely predictable. That it happened many times, in so many animal lineages, and so soon, would have been harder to foretell. The adaptability of the egg would turn out to be as important as the adaptability of the animal carried within it. But that is another story, of another time, for our next chapter.

4

STARBURSTS ON SHORES

*Ordovician Period, 488.3 million
years ago, to the end of the Silurian
Period, 419.2 million years ago*

'An nescis, mi fili, quantilla prudentia mundus regatur?'

('Do you not know, my son, with how little wisdom the world is governed?')

— Axel Oxenstierna (1583–1654)

Like soldiers carrying a battering ram, the legs of our Silurian millipede are coordinated, determined, unrelenting. Together, they carry the millipede's huge snake-like body across the Silurian land-dust. The millipede is larger and more rotund than it once was. Its body armour stretches at the seams, courtesy of the batch of fifty or more eggs it carries. It is looking for a place to lay them, but everything here is too hot. There is no forest canopy in this early land ecosystem and, hence, little shade. There are no leaves to fall and become a rich mulch. Instead, there is only mats and clumps of ground-hugging greenery through which the millipede has already trekked, eager to find the best spot. The female millipede pauses momentarily, its antennae gathering information on its environment – gauging the concentration of water molecules in its surroundings, measuring humidity, feeling the throughflow of drying air. Up here, on the dry Silurian continents, these things matter very much to invertebrates, still shaking free from their watery ancestry. There must be moisture – but where?

In the distance, there is the sound of a rushing stream, waves gouging at rock faces, grinding them to dust. Wet, glossy lichens cling to rocks, accumulating at their base a thin layer of something like mud. Upon finding this, the millipede readies itself to lay eggs. With its legs, it makes a moist cleft that will stay damp in the days that follow. Wet, white and oily, the eggs tumble against one another into a loose pile within the cleft, which the millipede arranges into a tight cluster. Using its antennae, it gently touches the face of each egg, almost as if counting them. It notes their consistency, their smell, their taste. Carefully, when it has finished laying, it treads soil upon the eggs, concealing them from their surroundings. Protecting them from elements – wind, drought, rain – common to land but unfamiliar to animals of the ocean. This was life on land back then, roughly 450 million years ago. A desolate arena for existence; a desperate place for eggs.

The millipede was not alone here. There were other animals back then finding homes on land. Centipede-like scuttlers; creeping spider-like forms; animals on their way to becoming insects. Scan through the fossil record of the Ordovician and Silurian Periods and it is clear that millipedes were one of many invertebrates that were crossing the ocean/land divide at this time, moving from the safe, cushioning bosom of the ocean onto the bleak continents, smudged green by low-growing mossy plants on the fringes. The arrival of these animals on land had nothing to do with bravery or pioneering spirit or anything like that. Rather it was the stacking up of numbers, of trials, over thousands or millions of years, occurring upon the shoreline where land meets water.

To me, it is unsurprising that so many sea creatures evolved to move onto the continents at that time. The land was, of course, an environment rich in potential in these early periods of life – filled with possible nutrients, lacking predators, brimming with potential for eggs. Yet I also understand why some people find it all so fantastical a concept to imagine that ocean or freshwater animals like invertebrates, and later fish, could take to an environment so alien to them. I mean, think about it. That water-living organisms can pull themselves from shorelines? Evolve muscular legs from fins and flippers? And have gills that turn into lungs? It really does sound somehow magical. But there is a certain inevitability to this evolutionary avenue.

To convey this point to students that struggle with the concept of sea creatures taking to land, I now point them to a simple demonstration that I came upon a few years ago from an experimental set-up created by Harvard Medical School in 2016, demonstrating how bacteria can evolve to live in a world surrounded by antibiotics at intense concentrations within only a matter of days (and hence why the overuse of antibiotics is a very bad thing). This example is a useful touchstone in explaining how and why it was always only a matter of time before animals evolved the ability to move onto land.

First, in the experiment, the medical scientists prepared a giant, rectangular petri dish, divided into subsections containing different concentrations of antibiotic solution. The subsection at one end contained agar with no antibiotics added; the next contained a little antibiotic solution; then ten times as much; one hundred times as much; and the final subsection was laced with one thousand times the concentration. Surely no bacteria could ever thrive at this upper extreme? Suffice to say, they can.

The results are both predictable and somehow simplistically, stunningly, beautiful. *E. coli* is introduced to the first chamber, the one without antibiotics, and they begin to proliferate. First an unseemly dot and then a blob; then collections of blobs; and then a bustling, thriving, overflowing community of bacteria. Within days, the first section is white as snow, full to bursting with populations of *E. coli*. At this point the neighbouring section, the one laced with a small concentration of antibiotics, is empty; the bacteria cannot cross into this antibiotic wasteland. But in the original section, millions of bacteria are still reproducing, until, suddenly, it happens: a single mutation appears – a single accidental triumph survives. A white dot appears in the next section and, in the hours that follow, spreads outwards into the new section like a spotlight beam. It resembles a starburst – a single flame licking at an untapped realm. Soon, other small pinpricks of antibiotic-resistant strains begin to break through the barrier; they too expand into the second subsection as they consume the agar, free from competition. Within days, this second section becomes swollen with populations of white bacteria, just as the first did. A few more days and they've breached the third section, then the fourth. And then finally, after eleven days, the final enclosure is crossed. Life has found a way around these hostile antibiotics and filled every available space on the agar.

A similar process took place with ocean animals, within just 100 million years of the Cambrian Period. First, a tiny explosion of life where the waves break; then in the surf, then the swash zone, exposed at low tide. Then the foreshore. Then the backshore. And then, finally, the land itself. Each organism a starburst into a new, hostile realm; the land their agar.

Once, at some point early in the history of animals, the millipedes went through these stages. In fact, it is widely thought that they were among the first of the land-living animal tribes.

The dusty, rocky continents brought three challenges to early land-walkers which, in the Silurian Period, starburst after starburst, millipedes and others evolved to tackle. The first challenge was how, without a fluid medium, sperm could meet egg and vice versa. The second challenge was how, in a world of chaotic seasonal climatic shifts, land-walkers could avoid laying eggs at the wrong time. The third challenge was the hardest of all: how to protect eggs from the ravages of Earth's bone-dry atmosphere.

The first challenge seems peculiarly, almost comedically, insurmountable. In the sea, sperm can be left to find its own way through the water to an egg. It swims competently and energetically like a tadpole, using (in most species) a long whip-like tail. But out of water, sperm are unable to move and, being diminutive and thread-like, don't survive for long. Over time, Silurian land-walkers evolved a range of simple solutions to get the two cell types to meet, some of which are still used to this day.

In prehistoric millipedes (and modern-day species) special male appendages evolved for the purpose of cradling and transmitting blobs of sperm to females. These appendages, known as gonopods, were used by male millipedes to pluck beads of sperm from out of the body, which the males then attempted to get into the reproductive tract of nearby females. In their union, male and female millipede faced one another and, after a few moments of careful prospecting through the mutual tapping of antennae, the gonopod was introduced to help sperm and egg become one. This sexual behaviour is broadly replicated in modern millipedes. Millipede sperms lack tails. In essence, they resemble eggs. Thus, not being able to swim, their sperm need to be scooped up by the gonopods and manually inserted into the female's reproductive tract. For this reason, in the parlance of millipede experts, gonopods are either 'charged', meaning their gonopods contain sperm, or 'uncharged', when they are empty. Collectively, these sperm blobs have a name – they are called 'spermatophores'.

Other spermatophore-carrying invertebrates walked the Silurian silts at the same time as early millipedes. In some fossils, their footprints are clearly visible. Commonly known as velvet worms, these early land animals resemble very closely some sea creatures in Cambrian fossil strata, although their appearance in the fossil record is fleeting, due to their being almost completely soft except for clawed limbs and robust mouthparts. These early land-walkers also hit upon a solution to the problem of how sperm and egg can meet away from the water. Although velvet worms resemble caterpillars or worms, it is the legs that set them apart – they move deftly upon numerous rows of pudgy, water-balloon-like pillars. As with millipedes, sticky sperm blobs are a common feature across the group. Today, there are male velvet worms which glue the blobs to the bodies of females, who digest their own skin to make a hole through which sperm can travel into the body. Once through the external boundary of the skin, the sperm swim through the internal body space of the female to the ovaries. It's possible that this may have been an adaptation first shaped in the Silurian Period, when animals like these ran aground.

Other land animals also began to modify their reproductive anatomy to better bridge the physical divide between sperm and egg.

Most notable among them are the sea scorpions (or 'eurypterids', as experts know them), some of whose probable descendants, the earliest arachnids (spiders, scorpions, mites and others) went on to festoon the land in what can only be described as hard-weathering sperm gardens. In the earliest arachnids, perhaps because eggs did not evolve such a hard shell, it was sperm, rather than eggs, that became the weatherproof form required to endure the dry, then probably very unsatisfactory for life, shorelines. Sperm hardened itself; eggs, in some animal lines at least, became very well hidden indeed, retreating further and further into the female body.

Like their sea-faring ancestors, as well as the millipedes and the velvet worms, the earliest arachnids deposited sperm in spermatophores for females to find and pick up. What these prehistoric spermatophores looked like is unknown, although modern descendants of these land-dwelling invertebrates offer some clues. Scorpion spermatophores, often glued to rocks or pebbles, are classically long and thin, almost the size of a toenail clipping. At their base is the pedicel, which glues the spermatophore to the substrate. Above this lies the stem, which contains soft membranes filled with spermatozoa, followed by the stalk – a hardened, ribbon-like structure, caramelised in its texture. Rather like a ring-pull opening, it is the movement of this stalk, when a female stands over it, that triggers the compression of the fluid in the chamber beneath. The sperm leaves its sleeping quarters through the 'capsule', which is adapted, depending on species, to direct sperm in just the right way towards the female's genital tract. Although modern scorpions have had more than 400 million years to adapt to the land, it's likely that they were performing a comparable reproductive ritual in their earliest days there.

Most mites in the modern day continue the primitive sexual behaviours of their arachnid ancestors by dotting spermatophores around their landscape, which females are prone to spotting on their travels. Under the microscope, these spermatophores resemble filament-like mushrooms held atop impossibly thin stalks. Their height differs between species, each having evolved to hang at exactly the right elevation to be noticed by, perhaps sampled by, a female of the correct species walking past. Within the construct of the mite's spermatophore is a small sac in which the waiting sperm sleeps, awaiting a 'kiss' to wake it from its slumber. Even a fleeting glance or a casual brushing-past by a mite of the opposite sex is sometimes all it takes.* Mites, being comparatively similar to the earliest arachnids, are the informed choice for how spermatophores may have been used in the heady days of the Silurian Period.

Within the Silurian fossil strata are the ghostly imprints of the ancestors of organisms like these, and sometimes it is hard to tell if they are made by spider-like arachnids or something else. Some trackways loop and ribbon around the slabs, representing land invertebrates in the process of measuring up chemical gradients, guiding them towards food. Other trackways are fleeting, partially washed away by routine tides or seasonal rains. Some are more linear, suggesting that they were made by land invertebrates with purpose:

---

* Spiders have evolved down a different path to scorpions and mites. They no longer produce the 'traditional' spermatophore. Instead, male spiders ejaculate onto a tiny web and then suck up the sperm using nozzle-like appendages near the face. Should a mating opportunity arise, males pump sperm via these nozzles into the female spider's reproductive openings (while trying to avoid being eaten).

running for shelter, fleeing from unknown predators, perhaps. Or guided towards . . . sperm packages? It's possible.

Crucially, through the formation of secure, protected spermatophores, males probably became more frugal with the number of sperm invested into each package. In many mite species today, spermatophores contain as few as 50 spermatozoa each and it may have been the same back then. The popular narrative has it that a sperm's chance of fertilising an egg is very slim – perhaps one in a million. For these prehistoric arachnid sperm, with their curious sperm packages, the chances were probably very good indeed provided the female found and accepted the spermatophore. As a result, the number of sperm in some spermatophores may have decreased with time, perhaps to as little as a few hundred in each package. And so, in the stirring Silurian climate, the characteristics of sperm and egg began to blur when placed on land. Sperm, almost, became more egg-like; it wasn't fruitful and cheap, the spermatophore became a more costly resource. And they were attached to substrates in much the same way eggs had been, underwater, in the Cambrian Period. It was almost as if the land flipped a switch in the evolutionary character of sperm and egg, re-framing and instilling in them new personalities, a new agency.

One can imagine them, the earliest arachnids, scouring at rock faces, picking through soils that were little more than aggregations of dust, tending to microscopic crops of spermatophores, like apple-growers in an orchard. It is more than a little amusing to imagine a Silurian landscape peppered with packets of sperm. And so, the first challenge to eggs on land – how sperm could meet egg – was, loosely, solved.

The second challenge that the new land-walkers faced related to the seasons. The land is a tempestuous place for potential egg-layers, with dramatic changes in food availability and temperature. Timing, over all things, began to matter. And so, another adaptation began to evolve in the Silurian. By the late Silurian, evolution saw to it that the early arachnids, millipedes and velvet worms were sharpening their season-spotting talents, timing their reproduction for when environments were at their richest and least uncertain. These animals stored sperm, putting it on ice (so to speak) for fertilising eggs later in the season when conditions were more welcoming for the laying of eggs.

Evidence for this comes from fossil eurypterids (as mentioned earlier, often called sea scorpions) from which arachnids sprouted. As with millipedes, fossils of eurypterids show sex differences between individuals of the same species. These differences are found hidden on the eighth body segment, which has an opening on the underside known as the genital operculum. Fossilised female sea scorpions often show evidence of a connected structure known as the 'horn organ', which is closely analogous to spermathecae – the organs for storing sperm found in female mites, scorpions, spiders and 'harvestmen' (Opiliones) in the modern day. Spermathecae are, essentially, fluid-filled internal sacs, sperm reservoirs for want of a better description, connected via a tube to the oviduct ('egg tube') where eggs are stored. By flushing this reservoir of sperm after ovulation, when the eggs tumble into the oviduct, the female lays batches of eggs already fertilised by sperm. Thus, through the development of spermathecae, sperm became something females could store and use to fertilise eggs at exactly the right time.

For land-walkers back then, among the first organisms to chart dry continents, the benefits to this adaptation are not hard to imagine. Compared to water, which heats and cools slowly, the land can change very quickly indeed. Heatwaves, floods, droughts, mudslides, storms – these were ever-present threats to life on land, just as they are today. Even the appearance of hard frost or dense morning dew is enough to incapacitate some small invertebrates. Individuals in the Silurian best able to spot seasonal change and modify their egg-laying behaviours dodged the chiselling forces of natural selection most effectively. And so, probably, during these early periods of life on land, invertebrates evolved improved sensory systems to read signals in the environment, particularly those relating to seasons. The earliest land-walkers developed their sky-watching systems – measuring changing day length, day by day, to determine (unconsciously) when the time was most right for laying eggs. These systems are poorly understood in arachnids today, but they are there. Red spider mites kept in a glass tank underneath a windowsill, for instance, will move through their life cycle regardless of the temperature of the laboratory. What they see through the window, how much light they are given, is what matters to them. These daily and seasonal rhythms exist in millipedes too, even those that spend their whole lives living in relatively dark caves. It is likely that these early land-walkers settled into these routines fairly quickly, perhaps within a few million years, or even sooner, modifying pre-existing neural routines evolved by their ocean-dwelling ancestors.

And so, to the final challenge laid down in the Silurian: how to stop a burning star and a parched, mostly waterless, relatively wafer-thin atmosphere, from killing a once-waterborne egg.

The central problem facing land eggs is that embryos need oxygen and water for sustained growth — in fact, embryos need plenty of both. The embryo needs more oxygen, particularly, than could ever be contained in an egg when first laid and so the exterior of the land egg has to be permeable to allow it access to new oxygen as it grows. To allow for this, in eggs of all species, big or small, invertebrate or otherwise, microscopic pores in the external membranes of the egg allow oxygen to flow in and waste gases (carbon dioxide) to escape from out of it. But the requirement for the egg to have pores means that water molecules can also evaporate from the egg, potentially drying the embryo out. A dry egg is a very bad thing for most animals. When eggs are exposed to air, often for days at a time, the moisture inside the eggs evaporates into the surrounding atmosphere, driven by diffusion, where water molecules move from areas of higher concentration (inside the egg) to areas of lower concentration (the environment that surrounds the egg).

The rate at which eggs lose moisture depends on the relative humidity of the surrounding environment and the temperature. This means that eggs laid in dry places with low humidity will have significant water loss, dramatically reducing the survival chances of the embryo. As a result, the best places to lay eggs on land are nearly always locations, microhabitats really, that are locally very humid. Another thing that can kill an egg, related to the above, is wind. Egg-laying locations tend to be away from airstreams because moving air disrupts concentrated 'clouds' of water molecules in the air immediately surrounding the egg, reducing the egg's local humidity even more, leading to more water loss for the embryo.

There is one primitive group of arachnids alive today that hints at how the earliest arachnids may have protected their eggs from these environmental threats. It is a sparse group of organisms known as hooded tickspiders, found in isolated parts of west central Africa and the Americas, that are unerringly neither one thing nor the other. Although they walk upon eight legs, hooded tickspiders move slowly, with slightly less certainty than 'true' spiders. Their bodies are shield-like and their armour unusually thick, almost crustacean-like. In fact, there are many that argue that hooded tickspiders are true 'living fossils' – descendants of those earliest Silurian arachnids which, beguilingly, have managed to hang on in some forgotten niche relatively unchanged.

The 'hood' in hooded tickspider refers to a shield-like roof on its dorsal side that can be pulled backwards to expose the organs beneath, including the mouthparts and the opening of its digestive tract. This protective veil also doubles up as a nursery for eggs and offspring, which the female hooded tickspider does all it can to protect, carrying them around from place to place until they are old enough to fend for themselves. It is possible this is how the earliest arachnids protected their own eggs and young from the ravages of our planet's atmosphere.

Another strategy employed by Silurian land-walkers to keep their eggs damp was to store the eggs in the humid confines of the oviduct (egg tube) for as long as possible, a practice many arachnids continue to this day. In scorpions, for instance, the young often hatch from eggs within the body of the female and are 'birthed' live. In mites (of many species) hatchlings stay inside the female's body for comparatively longer, sometimes eating the mother alive.* Millipedes, as we have seen, mostly favour damp soils.

These animals continue to navigate the challenges of our planet's drying atmosphere in their egg laying today. Depending on where you are reading this, there are likely to be hundreds of mite eggs in the damp cracks between slabs on your patio floor or, often, in the soil that nurtures your houseplants. In flowerbeds and beneath leaf-falls in local parks and gardens, there will be thousands of millipede eggs, deposited and then trussed up in their water-imbibed egg beds. These modern-day representatives of the Silurian eke out an existence among us today, successfully navigating the challenges to eggs that the land laid down back then. Once, they lived in the oceans. Then, they were drawn onto land.

The rocky continents in which our lives play out have a rich history that began long before our kind were here. Through hundreds of explosions on the shore, in successive waves of one in a million, this was their world before it was ours. And so, by the end of the Silurian Period, the animal egg was a biological ephemera within sea *and* upon land. On terra firma, a vessel that could weather an un-nurturing atmosphere; that could be timed to perfection; that could be united with sperm in new and surprising ways away from water.

This was just the beginning.

* There is even one group of mites where hatchlings have sex with one another in the confines of the mother's body — resulting in pregnancy within pregnancy.

Interlude
A post-Ordovician moment

The Earth enters a momentary blip in geological history. The atmosphere becomes temporarily cooler. The ice caps swell, the oceans retreat and the drip, drip, drip of ecosystem collapse begins. Sea levels fall gradually, just a millimetre or so every few months, but that is enough, year on year, for the coastlines to alter and re-form themselves on this late-Ordovician coral reef.

Now, twice a day, pulled to and fro by an orbiting moon, what was once a busy undersea community is regularly exposed to the elements. For hours at a time, wind and rain works the jagged, twisted surfaces of the uppermost parts of the reef. Like chimneys, horn-corals project upwards from the canopy, their tentacles tickling the wave tips and retreating like a closed fist when the waves fall too far. Upon them, when the tide recedes even further, tentacled anemones withdraw into aggregations of slimy blobs that look like puffballs on tree trunks. Yet, held upwards to the sky, temporary pools of seawater remain in some tiny gaps and crevices on the exposed reef.

There, mere moments from death, lying in one shallow pool and exposed to the drying wind, lies a shelled, octopus-like creature known as an ammonite. Its large, spherical eyes roll upwards into its shell. Its tentacles dangle lifeless. Rain rattles against its thick armoured casing and water collects in its whorls. And there, hanging beneath its body is a slimy package of long, pendulous eggs, glued tightly into a bunch, like translucent grapes.

Within an hour of being exposed to the air, the egg mass is already dying. Moulds and other fungi work the surfaces of each vessel, feasting on the nutritious jelly contained within. Then come other organisms. Tiny flatworms emerge from the moist confines of the exposed coral and, bathed in rainfall, begin to explore each slimy capsule. They glide upon the surface of the haul, exploring the geometry of the mass, as if testing for weak points. Sea snails are next. Their sandpaper-like tongues scrape through the jelly, eager to feast on the embryonic spoils beneath. Then, of all things, a fish. Two fish, three fish. Not large. Small and nimble. They shimmy up the banks of the coral using their pectoral fins. As long as their gills remain wet, they take oxygen in from their surroundings, and can survive in stormy weather very ably. When they reach the jelly mass, these simple fish rasp their jaws against the quivering surface, inching closer to the most nutritious bit: the dividing cells in the centre of each blob.

Hours pass. Heavy clouds move across the sky. The sun emerges briefly. The moon returns, pulling the sea back over the reef like a duvet on a bed.

5

A TALE OF TWO FISHES

*Devonian Period, 419.2 million years
ago to 358.9 million years ago*

'If ... we say that each human individual develops from an egg, the only answer, even of most so-called educated men, will be an incredulous smile; if we show them the series of embryonic forms developed from this human egg, their doubt will, as a rule, change into disgust.'

— Ernst Haeckel (1834–1919)

n our story of the egg, a single fossil discovered in the Gogo Formation of Western Australia has special significance. Found by a team in 2005 from Melbourne Museum, led by palaeontologist John A. Long, the fossil was immediately recognised as belonging to a placoderm, a widespread prehistoric group of fish covered in articulated bony plates. About 30 centimetres long, the fossil was labelled and boxed up in the field and then put aside for later study in the museum. Its significance was not immediately revealed. In palaeontology, a substantial find takes the right specialist at the right time, sometimes in exactly the right frame of mind, to notice something novel or strange. To Long and colleagues, the fossilised armoured placoderm fish, when it was unpacked and investigated later, presented itself perfectly.

The first preparation of this fossil revealed a delicate arrangement of fine bones, along with well-preserved larger elements, including the fish's head and braincase, which showed it to be a new species of ptyctodontid placoderm – one of a group of placoderm fish with large crushing plates for teeth, relatively rare in the fossil record. It was certainly novel. The fossil was deemed useful at this point, since it could clearly shed light on the anatomy and taxonomic position of the group among other placoderms. The 'that's strange' moment (a tell-tale foreshadowing of any important discovery) came, however, when the fossil was immersed for a few rounds in a bath of acetic acid, a solution not much stronger than vinegar which, carefully applied, removes the limestone matrix around a given fossil specimen. This solution, in time, would turn out to reveal more than just bones.

In the trunk region of the prehistoric fish, just before the tail, were some tiny fragments of bones that were so thin they were almost translucent. Among them a tiny set of articulated jaws was clearly visible, suggesting that this was a tiny fossil fish somehow situated within the body of a large fossil fish. It was a fish embryo, preserved in its mother's body. But that's not where it ended, because below the fossilised skeleton of this placoderm embryo there was something else – a string-like structure, notched and twisted like twine: this was, the researchers realised, a 380-million-year-old umbilical cord – the earliest so far discovered. Strange doesn't quite sum it up: the discovery suggested that, within just over 150 million years of the Cambrian explosion, the eggs of some fish species were employing a new strategy. Embryos were, it appeared, wiring themselves to their mothers, partaking in new intra-uterine adventures. The research team named their discovery *Materpiscis* – this was their 'Mother Fish'.

Mother Fish was, and remains, one of the finest fossil discoveries ever made. It is the first of its kind to show the attachment between mother and offspring that we mammals (unrelated to these fish) know so well today. It confirms that, by the Devonian Period, the egg had evolved a new place to hide in the ocean; no longer only found in pouches, or glued to rocks, the egg now occupied a new, secure space, inside the body of the female fish, only a short distance from the ovary in which the egg was made.

After the *Materpiscis* discovery, others followed. A closely related species of placoderm fish, *Austroptyctodus gardineri*, also from the Gogo Formation, was shown to have inside it three unborn fossilised embryos. And then a third live-bearing placoderm was announced. It was a type of placoderm from a more distantly related group (the *Incisoscutum*) whose fossils are also known from the Gogo Formation. The presence of tiny fossilised fish in one fossil skeleton had previously been mis-identified as the fish's last meal. They weren't food, it was realised, but unborn offspring.

As news of these fossil discoveries flushed through the international news cycle, scientific illustrators began putting forward their interpretations for how they thought Mother Fish may have looked in life. It is hard not to look at their artistic visualisations without being swept up in the story. With their blunt heads and rotund middle parts, Mother Fish and their kind resemble stony-faced toads with long, sweeping muscular tails. Their strange body armour sees them occupy a middle ground between mineral and flesh. Some artistic interpretations of these livebirthers, used for museum galleries or in non-fiction books or articles, have the young, spotlessly clean, dangling passively from their mother via an umbilical cord still rooted in the egg tube. Others have the Mother Fish, framed from underneath and silhouetted against the ocean surface, giving birth in a manner almost like a whale, contorted, showing a hint of agony, with young expelled from the egg tube in a mist of uterine blood. The truth is probably somewhere in the middle.

How did such an adaptation arise in Mother Fish and its kind?

To answer this requires a brief digression into yolk. For, importantly, there would be no umbilical cord at all were it not for this life-giving substance – found in eggs across the animal kingdom – that is particularly easy to spot in the eggs of fish.

In biological terms, yolk, enveloped in its own sac within the egg, is a nutrient-rich soup that provides nourishment for the developing embryo. The distinctive colour of most yolks comes from special molecules within it ('carotenoids') that protect the egg from the sun's radiation. But this isn't all. The list of ingredients in yolk is staggering. The yolk sac can contain 90 per cent or so of the calcium, iron, phosphorus or zinc in the egg. Its molecules provide the raw architecture of cells, contributing to the base components of DNA strands. Without it the animal embryo would struggle to get through even a few rounds of cell division. It is precious and priceless.

The yolk sac is a temporary resource. As the embryo grows and enlarges, the sac empties until, eventually, it shrinks to become part of the intestinal wall. In fish, the yolk sac attaches to a special cavity in the embryo's midgut via a stalk. This specialised stalk is, in *Materpiscis* – this Mother Fish – the umbilical cord.

Anatomically, the set-up in *Materpiscis* embryos appears similar to what we see in some sharks today and thus sharks provide a potential guide to what may have happened back then.

First, inside the body, young *Materpiscis* hatched from its eggs and drained its supply of egg yolk. Then, after the emptying of the yolk sac, the embryo began to connect, via the withered umbilical cord, to the lining of the uterus. In those Devonian seas, the umbilical cord, once empty of yolk, evolved to secure new opportunities to sustain its embryo. Although we think of the umbilical cord as simple and hose-like, in these early fish it had an almost 'hairy' complexion. It was covered in what are known as 'appendiculae' – special finger-like lobes through which molecules can pass, specifically the energy-rich secretions pumped into the uterus by the mother to assist the embryo's growth. The surfaces of each lobe were probably very thin indeed, perhaps less than a hundredth of a millimetre thick. Across this boundary passed oxygen, sugars, amino acids and water coming into the embryo. Going the other way, probably, went urea, the embryo's waste.

In essence, *Materpiscis* evolved an umbilical cord with its own unique thirst for nutrition, grabbing at what the uterus could offer, rather than sticking only to yolk like most others of the time. It was probably the same arrangement for many placoderm fish.

Today, we know through fossils, including from the Gogo Formation of Western Australia, of more than 335 genera of placoderms with various forms and ways of life. Some, like *Lunaspis* and *Gemundina*, had flattened bodies and hid underneath the sand, watching for passing prey. Others, like *Coccosteus*, were powerful swimmers, with long tails and large, triangular fins. Most placoderm species were carnivores, hunting with powerful jaws covered not by teeth, but with bony plates that looked almost like beaks. The most famous, known as *Dunkleosteus*, was an 8-metre-long predator with razor-sharp jaw plates capable of slamming shut with a force of 5,000 Newtons. This gave it a bite strength more powerful than that of a tiger. Interestingly, on some placoderm fish, fossils reveal the presence of special bony fins near the anal region, leading some researchers to suggest that they may have been used, in males, to move sperm into the female's reproductive tract, much as sharks and rays do today with modified fins known as claspers. If true, it may be that *Materpiscis* was one of a collection of livebirthers in those ancient seas.

What would the daily life of a heavily pregnant placoderm have been like?

By the Devonian Period, ocean ecosystems were gaining tier upon tier of complexity. There was far more of a high-octane energy to ocean ecosystems compared to the early days of the Cambrian or Silurian Periods. On the coastlines and upon the reefs, visible in fossils from the Gogo Formation particularly, the three-dimensional structure of this environment started to change. Corals were like towers now; branching, tangling, arching. Upon this bustling architecture were rafts of sea lilies (relatives of sea urchins and starfish), their feathery tentacles caressing every tide, flicking at every ripple. Devonian reefs like these became a maze of pits and crannies, caves and labyrinthine chambers, in which fish, trilobites and other crustaceans found safe retreat. Floating nearby, ammonites the size of tractor tyres, their wise-seeming eyes fixed to the reef walls, worked their tentacles through the gaps. Only the unmistakable silhouettes of large placoderm fish, their jaws sharp as gardening sheers, would have caused the ammonites to shake from their relentless, silent interest in potential food.

It was in this environment that our pregnant Mother Fish and its kind swam. Not all placoderms stayed near the reef while gestating their young. By measuring slight changes in day length, and by measuring the changing gradients of silt and mud in the water, many placoderms probably migrated in certain seasons, seeking sheltered coastal regions to give birth, including coastal lagoons and, perhaps, estuaries, as many marine fish do today. Although energy-intensive and dangerous, these migrations to shallow waters likely paid off in some (perhaps many) placoderm species by way of increased survival of offspring. Shallow water is harder for large predators to hunt in. It is also more likely to have provided plenty of hiding places, by way of seaweeds and overhanging vegetation.

One modern analogue to how some placoderms may have given birth can be seen in lemon sharks, a livebirthing species which in some populations migrates more than 650 kilometres to visit the same, exact lagoon each birthing cycle, a shallow mangrove creek in Bimini, a chain of westerly islands in the Bahamas. Here, the pregnant shark, almost as if in a trance, rests upon the shallow, sunlit sea floor among seaweeds. It eyes fix a stony glare and its mouth begins to rhythmically flex open and shut, ensuring enough oxygenated water flows across the gills. Deep within its body, contractions start. They are gradual at first and then, within minutes, gather to a crescendo. From beneath the female's body, a new life is ejected outwards. The first of many. Born backwards, the young shark's tail embeds itself into the sand and, as it lashes it from side to side for the first time, what results is a fine mist of mud and sand that stirs all around. Often, intentionally or otherwise, this conceals the birth of the next pups. The presence of these pups is exposed fleetingly in the dust-strewn mist that follows: momentary glimpses of fins and arched, slapping tails are all that can be seen. Throughout this ensuing drama, the adult female shark remains apparently stoic, not even glancing at the events taking place in the sand beneath its body. Sometimes six, sometimes, ten, occasionally up to eighteen pups are born. Attached to each one at the navel is the grey string-like umbilicus which ends in a bulbous blob – the hungry 'mouth' of the umbilicus which once attached to the female's uterus. Soon, as with the mammalian umbilicus, it will fall away and die. Once they have gathered themselves, the pups quickly swim off to the shade of nearby mangrove roots, a hiding place that they will frequent for several years before becoming big enough to see off the predators further out to sea. As breeding adults, ten or fifteen years later, many of the pups will return to the same Bimini lagoon, to have their own offspring. Provided they survive long enough, of course.

In the Devonian, Mother Fish and her kind probably feared sharks, then evolving alongside larger placoderm predators such as *Dunkleosteus*. These sharks were more torpedo-shaped than today, but in length they matched many modern-day species. Together, these predatory groups of fish were shaping the reproductive behaviours of medium and smaller placoderms, forcing the evolution of live birth (viviparity) and paving a way for some of the earliest animal migrations. Pregnant fish, journeying from the burgeoning coral reefs, swimming a gauntlet through shifting coastal sands towards the safety of shallow coves and inlets, made these worlds their own.

It is no surprise that, in some lineages, placoderms began to stray further and further away from the oceans during this period, many evolving into estuarine and river-dwelling forms. They were not alone in doing so. For other fish, more distantly related, were having some measured success in doing the same. They included among them, our ancestors.

In Devonian freshwater habitats, a different part of the fish family tree was also evolving to reap the benefits of occupying environments clear of large predators, rivals and competitors. It was near swamps and riversides that these fish were advancing to temporarily, clumsily, dip their fins out of water.

Just as the invertebrates had, in the Silurian and Ordovician Periods, evolved into land-walking forms, so too were early fish. It is probable that, in the history of the planet, thousands of fishes have evolutionarily flirted with the ecological boundary between land and water (both salty and fresh), each pushed and pulled according to their ecological needs and the communities of predators and prey with which each animal is interwoven. The Devonian Period, 419 to 359 million years ago, was, we think, when fish began to do so more frequently. Many of these early land-dabblers faced extinction, as nearly all things do, but the ancestors of one lineage, that would become amphibians, reptiles and mammals, would meet with spectacular success.

What would this rocky world have looked like to those first land-fish? From the water's edge, the plants in the Devonian Period were diversifying, pacesetting across land, as if laying out a carpet to encourage more animals from the water. Algal and bacterial mats, laid over red dusts devoid of organic matter, gave way to rich soil, a bed for terrestrial photosynthesis. The Devonian plants lacked obvious roots or leaves that you might recognise. They did not possess obvious collecting panels through which to gather light; their surfaces were more spear-like. In time, they evolved into lycophytes, early conifers and ferns. The earliest seed-bearing plants were also evolving here. So too, the structural plant architecture we call wood.

Throughout this period, when plants died there was still very little to break them down. And so, alongside bacteria and fungi, frequent warm rains were the land's deconstructing agent. Erosion is what gave life here. Down streams and rivers, the nutrients for life washed out of rock faces and into fresh water and oceans where they assimilated into the bodies of armoured placoderm fish, sea scorpions and trilobites. It was in this environment that the most successful landfish continued their evolution.

What were those landfish, our ancestors, like? And what were the evolutionary reasons behind this transition from water to land?

Tiktaalik, a late-Devonian fish discovered in Arctic Canada in 2004, is perhaps best placed to provide some answers, for it remains the best example we have of the transitionary stage that fish went through to make a life on land.

The famous fossil that details *Tiktaalik*'s upper half is exquisite. It portrays a large fish seemingly midway through its transition to land form. It has a triangular, flattened skull and a pair of fins, cleaver-like in shape. There is something crocodilian about *Tiktaalik*, with some individuals potentially reaching more than 2 metres in length. The fins upon which it propped itself had within them thick, strong-looking bones. It was a sturdy animal, easily strong enough to counter the terrestrial forces of gravity, albeit momentarily. Supported on its paddles, this was a fish that could survey its terrestrial space in a more efficient manner than almost anything else alive. Close analysis of fossil specimens of *Tiktaalik* suggests that the fish possessed spiracles, most likely connected to primitive lungs that worked in association with gills. *Tiktaalik* also lacked bony plates in the gill region of its body, meaning that it could move its head around: it is likely to have been one of the first vertebrates with a neck. It could move its eyes to focus on objects all around, from prey to potential mates, unencumbered by the bulk of its large body. It had wrists too. And the makings of shoulder blades. Subsequent fossil discoveries suggest that *Tiktaalik*, a resident of shallow pools and swamps, may even have been able to move from waterhole to waterhole over short distances, sloshing its way upon dry land. Were these water holes untapped resources for food or were they safe-houses for eggs? The fossils, sadly, tell us nothing. It could be one reason or the other. Or both.

Modern-day landfishes (species that have evolved to chart the mid-spaces between land and water today) help us imagine the ways that movement onto land can favour the survival of eggs. One note-worthy example exists in the fish group known as mudskippers.

To all intents, mudskippers look like your childhood image of a fish, but just a little saggy around the edges, as if they have been moulded out of clay and set in a kiln at too low a temperature. The mudskipper's pectoral fins are strong and muscular, easily capable of momentarily lifting the bulky body forwards in a manner not dissimilar to *Tiktaalik*. Although you'd be forgiven for thinking them clumsy, mudskippers are adept hunters, capable of catch-ing and eating insects and small crustaceans. Their eyes, firmly pinched upon the top of the head, offer wrap-around vision. Their gloss-covered lips and wet mouth (which allow for the transfer of oxygen directly into the bloodstream) give it a firm and serious expression. And, apparently independently, they evolved eyelids. Like us, mudskippers can blink to keep their eyes moist and 'cry' to remove sand and dust blowing around in their mangrove or shoreline habitats.

Key to the survival of mudskippers during their terrestrial periods is the complex burrows in which their eggs are laid. Among the designs that mudskippers employ are J-shaped burrows with a single entrance, U-shaped burrows with two openings, and W-shaped burrows that have three openings. Each design features a large, bulbous chamber in which the mudskipper will keep its eggs. This is a fish that has removed its eggs, mostly, from water.

Mudskipper eggs, often less than a millimetre in length, are more sticky than the eggs of other fish. The female mudskipper glues them onto the ceiling of an air-filled pocket in the tunnel, sometimes a metre or more below the surface of the sand. It is the male that will fertilise the eggs and then tend to them from this point, while the female goes off to replenish its reserves. When the tide sweeps away, and the noise of rushing waves recedes, the male remains in this claustrophobic coffin-like construction, with no way to turn around. But, for the mudskipper, nesting in such a place offers it something the hectic oceans cannot provide: a deliciously predictable safe haven from egg predators.

Today, mudskipper egg predators include crabs, sea snakes and skulking rock-pool fish known as blennies. Back in the Devonian, egg predators were almost certainly just as numerous, including sea scorpions, trilobites and perhaps the shelled octopus-like ammonites. Each predator – guided by scent molecules and, once close, sight – would almost certainly have been capable of making short work of fish eggs. And so, the land, a place mostly lacking animals with razor-sharp jaws or venomous tentacles, became a potential safehouse for the eggs of some fish species. Eggs were not the sole reason for fish to evolve to walk, of course. A particularly important factor behind the phenomenon was that the land was relatively untapped by way of resources – food, in the form of invertebrates, was probably plentiful. But predator-free pools and puddles were surely a favourable egg-laying option for some fish which natural selection could have readily, and predictably, exploited.

Today, there are other fish that experiment with land, and for whom eggs may be part of the evolutionary advance away from water. For rock-pool fish like the Pacific leaping blenny (*Alticus arnoldorum*), egg laying is a pursuit that occurs in the moist pockets and crannies of limestone boulders and cliffs, safe from the threat of predators. This species is so land-adapted that, it is said, specimens actively recoil when placed in water. Other examples include grunion fish, whose eggs are splashed onto wet Californian sands at high tide, deposited each year in their millions on moonlit spring nights. There are also the splash tetras, who leap from water to glue eggs on overhanging leaves and twigs. And South American killifish, some of whom lay eggs on waterside logs and branches when waters recede. The driving force behind many of these adaptations is undoubtedly predators. If this can influence fish behaviour in the modern day, why not back then, during the Devonian?

Fossil trackways of landfish, trails of footsteps left in hardened muds and sands, tell us more about how their evolution proceeded in the later stages of this geological period. These show paired impressions left behind from strengthened bony fin rods that resemble finger or toe bones, sometimes five strong, as they remain in us today. In a very real way, these were footprint-making fish. And so, in the Devonian, these early landfish managed something that sharks and placoderms did not. They left indelible prints in the fossil record that have continued, in various shapes and styles, to this day.

In time, the landfish and the sharks would go on to do rather well for themselves, but for placoderms, their demise at the end of the Devonian Period was relatively swift and devastatingly complete. Documented in the Gogo Formation, we see the diversity of placoderms decline and their ecological dominance subside in the period's latter stages. Quite why it happened is not entirely known but it is almost certainly to do with events at the end of this period, specifically the so-called Kellwasser event, around 372 million years ago, during which almost a fifth of all animal family groups faced extinction. The cause of this strange blip in the history of life is still much discussed. Some argue that the cause of the event was undersea volcanic turbulence, others that it was due to global cooling or even a meteorite impact like that which brought the reign of dinosaurs to an end. It is clear that life in the oceans appeared to face the brunt of an extinction wave, hinting at an intense period of water de-oxygenation that must have pushed the physiology of many marine and freshwater organisms to the brink. This would have been a fortunate moment in time to have been a landfish.

Right at the end of the Devonian Period, when so many ocean animals were drifting to extinction, while the placoderm fish died and the reefs bleached, these strange fish stood (relatively) tall, walking, striding in their own clumsy way, forwards to places no bony organism had moved before. In time, these organisms, their bodies ripe with eggs, would move across continents. Some would lose their bond, almost totally, with water, producing eggs able to weather the driest regions of our planet. And some, in the millions of years that followed, would reprise the role of Mother Fish all over again, this time on the land.

6

A MOST MARVELLOUS INVENTION

*Carboniferous Period, 358.9 million
years ago to 298.9 million years ago*

'Nothing is rich but the inexhaustible wealth of nature. She shows us only surfaces, but she is a million fathoms deep.'

— Ralph Waldo Emerson (1803–1882)

In a humid Carboniferous forest, new life was stirring in the form of insects. Among fallen, lichen-covered stumps these organisms skittered and danced. With many-lensed eyes, they watched from cracks in the nooks and roots of primitive trees. With antennae, they sampled the air, assessing the seasons, the environment, the proximity of local invertebrates that shared their spaces. Many of these early insects had unique appendages on the back of their bodies that other land animals of the time did not possess – they were long, protuberant scales attached to muscles that could be activated and flexed rhythmically. These were the earliest wings on Earth. And so, for the first time in the history of our planet, things routinely buzzed or chittered through the air. Under tiny gaps in the canopy, where shards of light danced and rippled across the forest floor, the early insects would have flashed in and out of our gaze.

In shape and size, the early insects, once little more than shoreline crustaceans, were becoming something else. Ecologically, they were evolving into a force to be reckoned with.

Many insects of this time resembled cockroaches, their waxy cuticles reflecting the canopy branches, still and silent. There were insects that resembled dragonflies of today. Although their flight was relatively cumbersome, some were as large as falcons. In straight lines, these gargantuan insects charted a path through the forest, assessing the activity of puddles and pools forming in the pits of fallen, fungi-covered trees. There were insects that resembled mayflies near these freshwater pools, their primitive wings held across their body like a pair of carefully ironed, translucent cloaks. There were some insects that were shield-like, flirting with a shape one might call beetle-like. All around, robust, almost robotic, forms that resembled grasshoppers or crickets achieved something that few land-living invertebrates had yet managed: they could eat leaves. In a freak twist many millions of years in the making, some Carboniferous insects stumbled upon digestive enzymes able to break down the tough proteins found in plants.

Soil, twigs, tree-top perches, puddles, streams – all of these microhabitats were becoming pliable real estate to these rapidly evolving invertebrates. In this warm, humid Carboniferous ecosystem, the insect success story was being crafted and cemented.

In part, the success of these insects during this period was courtesy of a new egg – another true land egg, with an armour-like covering far superior to previous insect versions. The evolution of the so-called serosa was occurring here – a wrap-around, breathable, security blanket for the insect embryo.

To get a feel for how innovative this new insect egg was, we must look to the previous incarnation of the insect egg, a far less waterproof vessel laid by insects from a simpler time. Before the Carboniferous, insects had been something of a collective of diminutive, scuttling six-legged invertebrates dodging a living among early millipedes and arachnids. The closest modern-day representatives of this line are the 'silverfish' found in homes across the world, and for whom this form of 'lesser' egg remains *de rigueur*.

Silverfish have no wings and their eyes are small and simple. There is no penis-like organ for males to transfer their sperm to females. Instead, male silverfish produce spermatophores, just like many arachnids and millipedes, which they lay externally. The female then uses spermatophores to fertilise eggs before secreting them in the damp surroundings they need for their growth to continue. In the past, these damp surroundings included humid Carboniferous forests, but now, in our homes, egg-laying sites are mostly restricted to condensation-covered bathrooms or damp kitchen cupboards. These are pretty much the only places that silverfish can lay their eggs because their eggs lack what the 'new' insect eggs of the Carboniferous developed: a water-retaining covering for the embryo – the serosa.

The serosa is a cellular membrane that almost completely enfolds the insect embryo in the early stages of growth, before secreting a cuticle enriched with chitin, a fibrous substance most commonly found in the insect exoskeleton. This protective boundary acts like a thin layer of concrete, slowing the rate at which water is lost from the egg in less-than-damp places. It is as close to waterproofing (and water-retaining) as a subject so small can be. So important is the serosa that, in the early stages of insect development, two thirds of all cells are invested in its creation.

The serosa, evolved in some dried-out bogland around 300 million years ago, endowed the egg with a longevity never before seen in insects. It allowed them to colonise new habitats and specialise in new microhabitats that no longer needed to be humid and wet. And, with it, insects, as an invertebrate group, seemed to flourish. Insects went from consisting of one or two families at the beginning of the Carboniferous to more than 100 families by the end. They managed that rare thing: evolving into new species more quickly than their competitors and facing extinction less frequently. This rapid evolution of new forms saw insects go from bit-part players running around in the shadows of early spiders and scorpions to, within only 60 million years, prominent cornerstone components of forest ecosystems.

What happened to the insect egg during this period rarely features in popular narratives about the rise of insects, which I find surprising. My hope is that the story can be remedied. Tweaked. Perhaps, re-told.

To understand the important role that the serosa plays in development, scientists have recently turned to the red flour beetle, a go-to species for genetic tinkering. Using a technique called RNA interference, scientists are able to manipulate the genes that code for serosa production, experimentally silencing them to see what becomes of the developing embryo. The results of experiments like these really are striking. Without the serosa, flour beetle eggs can still develop but only under one condition: high humidity. Switching off this gene turns insects into something more like their Carboniferous ancestors, the ones restricted to moist forests, such as the silverfish, whose eggs have no waterproofing.

The serosa is not the insect egg's only protective membrane. Almost like an undercoat, it is found underneath a hard layer that surrounds it, known as the insect chorion, which is most analogous to the hard eggshell that we see in birds. As with birds, this crystalline covering is added to the tiny egg as it moves through the oviduct (egg tube) by special nozzle-like glands in its lining. To add strength to the egg, the chorion is often sculpted with rings of hexagonal grooves or longitudinal ridges that can be seen using an electron microscope. Some insect eggs are so covered in bony protuberances that their surfaces look like coral or the shells of spiny sea urchins. Some are pitted like nectarine stones while others look sleek, metallic and streamlined like interstellar probes in a science fiction movie. Across the surface of the chorion, there are arrangements of microscopic pores ('aeropyles') through which oxygen and carbon dioxide can travel. A single opening, normally on the uppermost side of the egg, is the pore through which insect sperm enters before fertilisation. The chorion's strength comes from layers of lipoproteins, which also give insect eggs a wax coating – a kind of shiny finish. As well as offering protection, this waxy glaze helps insect eggs stick to surfaces, allowing adult insects to glue their eggs to leaves, rocks, branches or clumped together in soil. The number of pores on each egg, and their arrangement, differ between insect species, meaning each egg has not only its own texture, but also its own unique atmosphere.

The chorion provided structural support, but it is the serosa, particularly, which may have unlocked a new way of life for insects. With this improved weatherproof shell, insect eggs laid in what were once sub-optimal locations did not die as before. Instead, they thrived. Almost certainly, this saw insect populations change in the Carboniferous; it saw them spread, all the while adapting to new habitats and ecosystems as they went. In time, the serosa saw insects move further and further across the dry supercontinent known as Pangaea, which was newly gathering at this time in Earth's history.

Pangaea's formation was gradual, but the beginnings of its coalescence occurred during the Carboniferous Period. It was around 300 or so million years ago when the 'then' ancient continents of Laurasia (present-day Europe, Asia and North America) collided with Gondwana (present-day Africa, South America, Antarctica, Australia and India) and, later, Siberia, to create a sheet of land that covered approximately one third of the Earth's surface. The repercussions of this continental union were felt across every inch of its surface. Firstly, by fusing against one another, the Earth lost many of its most productive regions: coastlines. With their rich reef ecosystems and diverse communities of animals, many of these important ecological zones were simply pinched from existence. Secondly, in some parts of the world, the rate of erosion increased. And as the continents that made Pangaea continued to compress, great rippling mountain ranges were pushed skywards. These, in time, were washed into sands and muds, creating vast floodplains and deltas nearer the edges of landmasses that heaved with sediments. In equatorial regions they became dense forests, incorporating vast swathes of wetlands that were rich in life. But the mountain ranges have special significance for the climate of this period, for they effectively blocked moisture and rainfall from reaching the inner regions of Pangaea, causing it to become a vast dustbowl. Climate models, along with the study of fossil soils retrieved from the Moradi

Formation in northern Niger, suggest a region comparable with modern-day deserts like the Namib Desert of Africa or the Lake Eyre Basin of Australia – that is, an arid climate with short, sharp wet periods, often with catastrophic flooding. Dry habitats became more commonplace on Earth. It was into these places that insects may have spread, evolving as they went.

There are a number of familiar insect subgroups that have their deepest and most distant roots in the latter stages of the Carboniferous. They include the most diverse insects of all, beetles, which alone account for almost 400,000 insect species today. Other distant ancestors that started making a name during this time (or not long after) include fleas, scorpionflies, snakeflies, barkflies and the insects that would become grasshoppers. The diversity of these groups is testament to their nimble evolutionary talent for adaptation and survival. Insects are, evolutionarily, very hard to extinguish. And so, in the modern day, there are insects that endure in snow, that can travel higher than the Himalayas and that thrive in caves a kilometre or more below the Earth's surface. Today, insects pollinate our crops, they tidy our landscapes, they eat our waste. They are both small and large parts of the diet of 2 billion people each day. Our world would be unrecognisable, unliveable, if they were to suddenly disappear. Could eggs really be the reason for their evolutionary success back then?

The secret behind the rise of insects has long been argued – in books, in documentaries, in journals – to be about flight. The argument goes that insect wings, which evolved at some point in the middle of the Carboniferous, unlocked the potential of insects to move from resource to resource: that flight permitted the mixing of genes over a wider area, churning out more and more favourable forms to face the chaos and uncertainty of life. Insects evolved more readily, in other words, because of wings. This is almost certainly true. But, for me, there is an important addendum: that sturdy, weatherproof eggs are what is required for those genes to spread. Early insects could fly to dusty lichen-covered canyons; they could glide over desert oases and perch upon ferns and mosses; they could huddle under piles of rotting seaweeds and feed upon tiny mites and crustaceans. But, without a serosa, these insects could only have done this for fleeting periods: what matters are eggs. Eggs that can survive in these non-humid locales. The serosa paved the way for this change. And so, I consider eggs and wings as something of a tandem evolutionary display. A Carboniferous double act, perhaps.

In the modern day, even as you read these words, trillions of insects are going through the routine of egg laying and hatching, an action practised and honed in the hazy, drying days of the Carboniferous. The elegant placing of cabbage white eggs upon cabbage leaves, for instance. Bee-flies dropping their eggs like bombs into the nest holes of ants. Damselflies dancing like riverside nymphs, straddling the leaves that hang over the water's edge as they look for places to lay. Ladybirds, sampling the stems upon which aphids are most numerous, laying their bulbous eggs from which predatory offspring hatch.

Shield bugs, whose waxy eggs often reside in clusters on the undersides of leaves in spring and summer, are a personal favourite of mine. These eggs, glued in rows with surgical precision, are slightly cylindrical in shape, with a slight dentition at each end, almost as if given a squeeze with a pair of surgical tweezers. They are perfectly symmetrical, like intricately cut jewels waiting to be threaded onto a necklace. Within each egg, through the translucent window of the chorion, two broad circular smudges can clearly be seen. Upon closer inspection with a hand lens, these smudges gain definition. They are, in fact, the eyes of the tiny embryo, still within the egg, inspecting the world it will next inhabit. At the topmost side of each egg is a circular swelling, which will become the fault line upon which the egg will eventually crack. The eggs of birds hatch chaotically, every fragment of shell unpredictable in shape and size. In contrast, insect eggs crack open with conformity. When insect eggs hatch, it looks so clinical; machine-like; like a spaceship door opening. The lid neatly comes off many insect eggs, popped open by the pressure generated within the egg by the writhing of its occupant. By squirming leftwards and rightwards in a shuffling motion, the young shield bug leaves through this escape hatch and clambers onto the leaf, where it gathers itself. Here, the newly hatched insect changes quickly. Its hair-like antennae inflate, its chitin armour hardens, it twitches its legs as if proudly testing out electrical routines. It is ready to go. Every day, trillions of insects go through the routine of egg laying and hatching like this. It is an action secured by natural selection since the Carboniferous Period, when Pangaea first formed.

In some insect species today, low humidity has seen the serosa evolve to a near-Herculean degree, which can be problematic if their ecological needs happen to overlap with our own. Among the most infuriating to humans are the eggs of clothes moths, whose adult forms see in our carpets and other furnishings an expansive two-dimensional landscape of animal hide to feed the hungry mouths of larvae. Clothes moth eggs, each the size of a pinhead, are almost always too small to see. Hence, their entry into our furnished home happens without our even noticing. Without chemical treatments, these eggs remain hard to destroy, so much so that museum specimens infected with the eggs of clothes moths are often stored, temporarily, in freezers. One week at −18°C is normally enough to kill individual eggs. At −30°C it happens in seventy-two hours.

Many insects get their eggs into our houses through more specialised ways by using an intermediary. The eggs of head lice (which, although they sound crustacean are, in fact, insects) come in on children and make their home among the family. The 'nits' on their hair are, often, what we first notice. Nits are the millimetre-long egg cases glued onto individual human hairs, capped with a lid through which the embryo's air exchange occurs. So tough are these egg cases that, after the young louse hatches from them, the eggshell can remain in place for more than six months. Insect eggs like these are nothing if not resilient. The same is true of fleas, whose pearly eggshells gather in the corners of rooms, on the wooden struts under mattresses or in layers of sediment beneath the sofa cushions upon which the dog soundly sleeps. Again, there is that word: resilience. And let us not forget the eggs of bed bugs, whose egg removal requires weeks and weeks of sweeping with a stiff brush, hot temperature washes and, often, in frustrated defeat, a completely new bed and mattress. Many populations of bed bugs, and their eggs, have evolved resistance to almost all effective insecticides.

The difficulty we have in removing insect eggs like these from our homes tells us something about how insects came to take such a prominent role in the story of life on Earth. It is because they are steadfast, unthinkingly resolute in their 'intent': to guide the insect embryo on a journey towards sex, in a casing hardened to wind, to dryness, to the intemperate atmosphere of our planet. Theirs is a waterproofed world within a world, crafted in the days of the Carboniferous; among accumulations of liverworts and mosses; in the shade of fallen trunks and branches; among early arachnids, millipedes and centipedes evolutionarily wedded to a humid landscape from which the insect group was unpinning itself.

It all sounds so easy in the writing. Like the entire success of insects is some simple Just-So story. It was 'just so' that insects evolved hardier eggs and wings at around the same time in history. It was 'just so' that the weatherproof serosa evolved at exactly the time when Earth was becoming a drier, dustier place on land. But, I feel I should underline the point that there was (and is) no 'just so' to the story of how insects became so successful; there was no 'Goldilocks moment' for them that involved the lining-up of events in just the right way to make everything exactly right. It is *the events themselves* – the changing of climates, the movement of continents, the changing of ecosystems – that writes the guiding narrative here.

Our ancestors, now far more than mere landfish, were among other animal groups likewise evolving newer, tougher, hardier eggs, with relative gains and successes. Just as in insects, developments in the eggs of these terrestrial vertebrates were undoubtedly guided by climatic changes occurring at the time to do with Pangaea's formation. And this new egg has a name: the amniotic egg.

The animals that hatched from these revolutionary land eggs skittered among slimy tree roots, fallen branches and trunks, probably feeding mostly upon insects. Their jaws were filled with peg-like teeth and their eyes were sharp. They scuttled from shadow to shadow, with searching eyes, and an instinctual nervousness of their surroundings. These were distant descendants of the Devonian landfish, on an evolutionary journey to become animals we now think of as reptiles, birds and mammals. They would have resembled, in their fleeting moments, simple lizards. The 'amniotic' egg was their version of the serosa-imbibed insect egg.

To reiterate, it was no coincidence that the two new eggs started to evolve at roughly the same time. If variation exists in a population, if some eggs are even microscopically more resistant to drying than others, natural selection will, over time, be drawn to 'act' and populations will change. And so, the same force that saw fish take to land in the first place or, for that matter, sees bacteria evolve anti-bacterial resistance today, chiselled in the landfish and the insects a new kind of water-retaining vessel.

The amniotic egg was cooked up, evolutionarily speaking, under the surface of the ground rather than upon it; from eggs stuffed into the crooks of tree roots or within muds and moist soils. The lizard-like animals that laid them were precursors to modern reptiles and mammals, as well as a host of less successful 'tetrapod' (think: four-legged) groups whose extinctions litter the fossil record.

The secret to the success of the amniotic egg was its compart-mentation. These eggs are collections of fluid-filled membranes each with its own role. Four membranes are particularly important. The first is the amnion, which surrounds and protects the embryo. Filled with amniotic fluid, this provides a stable environment that cushions the embryo as it grows. The next membrane is the allantois, a hollow sac-like structure filled with clear fluid which provides a surface for gas diffusion to occur across, as well as for the removal of waste. (In amniotic eggs, waste gases build up in a little pocket of air visible at the axis of the egg – this gas pocket can be seen clearly, if you are so inclined, when taking the shell off a boiled chicken's egg.) The yolk sac, another easily observable feature in many amniotic-egg-layers, provides food for the developing embryo and can be seen to clearly shrink as the embryo grows in size. In birds, turtles and crocodilians, there is a further component, the albumen (or 'egg white') which plays a role in preventing microbial attack. Encompassing all these membranous sacs is the chorion, the enveloping armour that also assists the embryo with gas exchange. The entire construction of the egg is surrounded by another layer, the shell, which varies in thickness between species and is completely absent, of course, in nearly all modern mammals, including ourselves. The shell layer is pitted with pores through which gases can travel in and out. In the earliest amniotes (the name given to amniotic-egg-producing vertebrates), the shell was thin and flexible, completely unlike the crystalline eggs of birds and some dinosaurs (detailed in later chapters), which would take another 200 million years to evolve.

The arrangement and numbers of pores in non-mammalian amniotic eggs are shaped by the conditions in which they are laid and differ dramatically according to species. Developing embryos cannot yet use their lungs, so their breathing occurs through the simple diffusion of gases in and out of the shell. Each pore, a few thousandths of a millimetre in length for most non-mammalian amniotes, opens as a hole on the surface of the egg, attached to a 'pore canal' – the tunnel which leads to the egg's internal chamber. In many ways, the pores of amniotic eggs resemble the stomata of plant leaves or the spiracles through which insects breathe. They are nothing more than open holes, where the interchange between environments occurs. The egg breathes through them. To maximise the surface area available for gases to diffuse across, the embryo presses a network of blood vessels across part of the egg's internal surface, like a child's hand pawing at a shop window.

In modern amniotic-egg-laying animals, the number of pores on each egg varies between species according to their ecological niche. In birds (whose shells are the most well studied) there is a clear pattern: those birds that breed at higher altitude have fewer, smaller eggshell pores, a relationship governed by barometric pressure and the effect this has on water vapour. High altitudes see eggs dry more quickly, and so natural selection sees to it that the number of pores in eggs is reduced. That there are patterns between species, according to their altitude, is perhaps expected, but, incredibly, in studies of chickens, individual females appear able to modify the number of pores according to the altitude at which they lay eggs. Take a chicken to the top of a mountain and it will lay a different kind of egg to the one it would have laid at the bottom. Somehow, and scientists don't appear close to answering how, a bird's brain can measure atmospheric pressure and transmit this information to the reproductive tract in which its eggshell is being formed, modifying the number of pores the egg will contain.

How did this new land egg form? How did the amniotic egg evolve?

Answering these questions is hard, because the eggs of the land-fish descendants (tetrapods), being soft and quick to decompose, did not fossilise readily. Eggshell, whether soft (as it was then) or hard (as it became in birds) is, by its nature, fragile. It can be physically damaged or destroyed or washed away far more easily than bones. And, to top it all, calcium carbonate – a key component of many eggs – has a habit of dissolving in acidic environments. (Skeletons, composed mostly of calcium phosphate, do not.) We have very little direct evidence, in other words, to 'see', through fossils, of what happened to their eggs over time. But the general steps are popularly posited as follows.

First, the early tetrapods laid eggs with a jelly-like covering, akin to fish, from which they evolved. Over time, to decrease the rate of evaporation from the egg, the gelatinous covering would have been replaced with a fibrous boundary layer. This fibrous membrane provided the embryo with an added benefit: the egg could support a larger embryo without risk of collapsing upon itself. This permitted a larger, metabolically more active embryo to develop before hatching, unlocking a kind of ecological 'superiority' compared to their (mostly invertebrate) competitors. But size brought with it a new problem. For a larger embryo requires more by way of oxygen to grow, something the surface area of the egg is ill-equipped to support. In popular parlance, tetrapod embryos, evolving to larger and larger sizes, began writing cheques their eggs could not cash. And so, the tetrapods were constrained by their eggs – as adults, they remained relatively small: often less than 30 centimetres in length. To evolve to a larger size, gifting them ecological advantage over their rivals, it was the egg, rather than the body, that natural selection was unthinkingly drawn to work upon. The only way for the egg to become bigger was for internal structures in the egg, that aided respiration and the excretion of waste, to evolve – hence the impressive system of internal fluid-filled membranes that make the amniote egg so different to others.

The amniotic egg is, in evolutionary terms, more than just a terrestrial safehouse. It is more like a locked room, for single occupancy, inside which is a well-stocked kitchen, a comfortable bed, good insulation and a chemical toilet. Alfred Sherwood Romer, the noted palaeontologist and among the first to popularise the amniotic egg's evolutionary significance, called it 'the most marvellous, single invention in the whole history of vertebrate life' – an exposition that he went on to repeat for more than three decades, in textbooks, articles and works of popular science.

In the history of egg science, Romer plays an important role. Not only did he help crystallise the complex relationships of early amniotic-egg-layers and oversee an important time in Harvard's Museum of Comparative Zoology, Romer was also an experienced field palaeontologist, often working alone scouring the 'red beds' of Texas, laid down 275 million years ago. It was from here that he described what was then thought to be the first known fossil of a shelled amniotic egg, a 6 by 4 centimetre fragment, whose origins were later contested.

Romer's consistent view was that the evolution of the amniotic egg was 'epochal' and that its evolution unlocked the evolution of reptiles, birds and mammals, emancipating them from the watery world from which they had evolved. In more recent times, scientists have pointed out the old-school spin that Romer put on tetrapod evolution: that fish were simply half-formed amphibians; that amphibians were half-formed reptiles; that reptiles were half-formed mammals; that every animal on Earth was somehow on a journey aspiring to become us. The truth, as is clear, is more complicated – more muddy, in a philosophical but also literal sense.

The popular story of how the amniote (soft-shelled) egg evolved is often portrayed with a slightly Biblical bent. I remember being told, early in my academic studies, of prehistoric reptiles moving away from the wetlands ruled by amphibians and heading into an uncertain wilderness – exploring the driest, most open deserts where they then evolved shelled eggs with which they overthrew their rivals. But the soil, rather than the dry deserts, was almost certainly the bed of early amniotes. It is where we began. Scratching holes and digging burrows. We really are creatures of clay.

Many modern amniotes – especially reptiles – still use soil as a substrate in which to lay eggs. These include many monitor lizards, skinks, snapping turtles and snakes. The benefit is that soils take a long time to completely dry out. They remain moist for longer and this suits the needs of the amniotic egg, reducing the rate at which water is lost into the environment.

Modern-day tuataras – lizard-like reptiles from the islands of New Zealand – are especially interesting in this regard. The tuatara represents a lone surviving twig of a particularly ancient lineage of the amniote line that has somehow arrived, persisted, into the modern day. Tuatara eggs are covered in a soft, parchment-like covering just a fifth of a millimetre thick. Their egg burrows can reach 20 centimetres below ground. But females actively choose the kinds of soil that their eggs need to grow. In a scene reminiscent of what their Carboniferous relatives may have endured, some female tuataras leave their forest habitats to seek warmer, more open soils for egg laying. Presumably, these females are adept at 'sampling' soils as they migrate. The same is almost certainly true of all modern-day reptiles that lay their eggs in sands and soils. Rather like prospecting farmers, these animals recognise what is, and what is not, a good substrate for their 'seeds'.

But there now begins a twist in our tale. The evolution of the amniotic egg in the ancestors of reptiles, birds and mammals brought with it the requirement for an anatomical change in males. To be fertilised, each egg requires sperm to meet the egg before being covered by shell, which is produced by special glands in the female reproductive tract. This logistical requirement favoured males best able to get their sperm directly into the female tract and so, predictably, the evolution of the penis becomes a part of our evolutionary story.

Internal fertilisation in this way is not unique to reptiles and mammals. Far from it. The anatomy to deliver sperm directly into the female body has evolved numerous times among fish, including in sharks and rays and, of course, placoderm fish. You will remember that in sharks and rays, the male's sperm is guided towards its target through adapted fins – the claspers, mentioned earlier in our story. The origin of the penis in reptiles, birds and mammals has its own unlikely origin. This intromittent organ is actually a leg, of sorts. In 2014, scientists in Cliff Tabin's lab at Harvard were able to show, using lizard stem cells artificially labelled with fluorescence, that cell lineages that formed the penis were originally those used to create limbs. This explains why some amniotic-egg-layers, most notably lizards and snakes, possess not one but two penises (known as hemipenes). In snakes, the tissues originate from embryonic limb tissue. In mammals, which we visit later in our story, the origins of the penis are even more complex. It appears that, early in mammal evolution, the cluster of cells that become the penis transitioned away from using limb precursor cells to the precursor cells that produce the tail. Why this may have occurred is not yet known.*

Regardless, in the words of science writer Liam Drew, the amniotic egg is why we have a matrimonial bed and not a matrimonial bath. It is also why you have, perhaps nestling beneath this book, an umbilical scar – the sealed door that once, many years ago, connected you to a placenta, the membranous organ through which the vital molecules that made you were transferred – a modified structure which has its roots in the structure of the reptile-like amniotic egg from which the mammalian lineage began. But more on that later . . .

Upon evolving the chorion and serosa, a penis-like organ also evolved in male insects, retained in most species alive today – this is the *aedeagus* as entomologists know it (derived from the Ancient Greek *aidoia*, 'private parts' and *agos*, 'leader'). This intromittent structure, which is pressed into the female's reproductive tract in many insect species during sex, has its evolutionary origins in abdominal appendages once found upon two of the rear-most segments of the insect body plan. It works in approximately the same way as the penises of reptiles and mammals, although insect sperm is delivered in slimy blobs rather than aqueous drops.

As mentioned earlier, it interests me that both the ancestors of reptiles, birds and mammals (collectively termed amniotes) and the

---

* It is worth stating that most male birds have lost this anatomical endowment, perhaps to lighten the body load and permit flight. Instead, male birds transfer sperm through a 'cloacal kiss' between males and females. The ostrich, with a so-called pseudo-penis, is, in an evolutionary sense, an outlier.

ancestors of most modern-day insects hit upon the same kind of egg at roughly the same time in our story. The Carboniferous climate – dipping, drying, cooling as it passed – was almost certainly central to this. The suggestion is that these two unrelated land eggs were shaped by shifting continents, moving at the mercy of a violent, volcanic netherworld. And so, although it might sound strange, the hard-wearing land egg is an object shaped by a planet, a product of a brooding rock cycle that has churned on and on for billions of years.

Needless to say (placental and marsupial mammals not with-standing) the 'design' of both eggs largely stuck. Together, from the Carboniferous forests, the amniotic-egg-layers and the insects began to chart new ecological zones, planting eggs further and further from the humid and damp soils, seeding them in gradually more parched canyons and floodplains, uplands and lowlands, across previously inaccessible regions of continents. What a thing this would have been to see from above: insects and the ancestors of mammals, birds and reptiles spreading across our world for the first time; cementing lineages that would last for hundreds of millions of years, that we share our lives with today when we smile at every birdsong, stroke our favourite cat or dog, curse every nit and wince at the pearly egg of every flea.

As Romer declared, the amniotic egg really is a most marvellous invention in the history of life on our planet. But for me it was one of two incredible developments, united in a single common evolution-ary habit, hard to shake off and shared by all things. To make more. For as long as possible. To be everywhere.

7

THE LARVAL STORM

*Permian Period, 298.9 million years
ago to 251.9 million years ago*

'Your worm is your only emperor for diet; we fat all creatures else to fat us, and we fat ourselves for maggots.'

— *Hamlet*, Act Four, Part One

From space, the temporary conglomeration of continents known as Pangaea was like a large, brown stain upon Earth's surface. Each day, its oppressive face would rotate towards the sun and be cooked by it. Storms upon its edges would tumble and swirl, the rains they carried rarely penetrating its heart. A climate that began to change in the Carboniferous continued to rage. It became an increasingly arid world, particularly in its cracked, warped and scorched interior regions. But across coastlines, more temperate and humid climates prevailed; moist fringe lands greened with life, becoming a ring of productive ecosystems of conifer forests, with ginkgoes and cycads and seed ferns and a host of other new evolutionary plant innovations steadily gaining ground. In the rock faces of South Africa and South America we find evidence of some of the early reptiles that lived in these maturing Permian habitats. While these fossils tell us plenty about the diversifying reptiles that had evolved by this time (newt-sized procolophonids, with triangular turtle-like skulls for tearing leaves off branches; lizard-like millerettids, clearly insectivorous according to their diminutive, yet numerous, pointed teeth; the therocephalians, cat-sized predators with well defined, muscular jaws), almost nothing is known of their egg-laying habits. There are almost no fossils to guide us, their fragile remains lost to the hands of time. Occasionally, however, fleeting glimpses of amniotic eggs – the eggs laid by the ancestors of modern mammals and reptiles – have appeared in the Permian fossil record, visible only to field palaeontologists with the sharpest of senses and, back at the lab, access to the finest technologies.

Synchrotrons, powerful particle accelerators that fire electrons through solid rock, are one tool researchers now use to investigate potential fossils of amniote eggs entombed in a matrix. Another tool is the CT scanner, which sub-divides fossils into three-dimensional layers that can more easily be explored using computer software. It is through technologies like these that we can glimpse, albeit only in a handful of specimens, how amniote eggs were evolving in the Permian Period.

Two fossil egg discoveries in recent years are especially alluring. These fossil finds are both from Permian mesosaurs, freshwater reptiles that superficially resembled crocodiles and that occupied wetlands, including hypersaline lakes and ponds, 285 million years ago. At 1 metre in length, mesosaurs were competent, well-established predators by the standards of the time. Mesosaur (not to be confused with mosasaurs, the group of Cretaceous marine lizards) fossils have not only been discovered in South America but also in southern Africa, and for good reason – because, in the Permian, these two continents were connected. In fact, their appearance in the fossil record across both sides of the Atlantic was early evidence to support the idea of continental drift, more than a century ago.

These two Permian amniotic egg discoveries are spectacular. They show us, for the first time, details of amniote mesosaurs in their earliest and most vulnerable moments. And, importantly, there is the slightest suggestion that, within 50 million years of its evolution, the amniotic egg was becoming an internal egg – one held in the female's reproductive tract rather than one laid on land, in soils or sands. This was a Mother Fish, or something like it, resurrected somewhere new, in a different time in Earth's history.

The first fossil discovery, in 2009, was by palaeontologist Graciela Piñeiro of Uruguay's Facultad de Ciencias, who found what she assumed was a coprolite (fossilised faeces) in a pile of debris in an abandoned quarry. About 15 centimetres in length, Piñeiro was hopeful to find evidence of fish scales and bones in the coprolite. Instead, upon cleaning and preparing the fossil, she discovered that it was a tiny mesosaur, curled around itself in a tight ball. Piñeiro's original misdiagnosis of the fossil is understandable. To look at, the rock resembles a rag-tag assemblage of disarticulated bones, fragments intermeshed against one another with no obvious pattern. But hold your eye over this specimen and order begins to form. The fossilised bones glitter and shimmer and the obvious features of the amniote body plan begin to crystallise. Backbone. Ribs. Shoulder blades. Its long skull glares rightwards, filled with needle-like teeth amid precision jaws. Immediately below the jaws two diminutive limbs can be seen, curled upwards and cupped under the chin. Lastly, filling up the remaining space, a long and sweeping tail is pulled back on itself, draped across the belly and folded under the chin like a treasured blanket. The level of preservation is extraordinary in this fossil and this is no accident. The high level of detail is down to the fact this tiny mesosaur perished in conditions favourable for the preservation of fossils, namely stagnant, stinking muds, probably high in dissolved salts, that discouraged the decomposing actions of invertebrates and micro-organisms. In conditions like these, even nerve fibres and blood vessels of the fossilised embryo are observable. We know about mesosaurs, not because there were necessarily many of them, but because many of them died in exactly the right environment for fossilisation to occur.

Piñeiro's first thought upon realising the significance of this tiny mesosaur was that it was the remains of an embryo, laid on land, whose soft shell had been lost during the process of fossilisation. But then, surprisingly, a second mesosaur embryo was discovered, this time in Brazil. This embryo appeared to be inside the fossilised body of an adult mesosaur, suggesting that mesosaurs retained eggs inside the body, as many mammals and reptiles do today, including anacondas and many sea snakes. The ribs of the mesosaur embryos, in both specimens, are thick and robust, suggesting that young were able to swim confidently once they emerged.

Fossils like these show that the amniotic egg was changing in the Permian Period, continuing to adapt – concealing itself within the uterus of its mother for longer in some species.

Mesosaurs were, quite probably, the first reptiles to travel down this evolutionary path, but they would be far from the last. In later periods, other marine reptiles would also, convergently, come to tread a similar avenue, going from what may have been an egg-laying way of life to giving birth in a manner broadly similar (viewed from the outside at least) to whales today. These 'livebirthing' marine giants included dolphin-like ichthyosaurs and long-necked plesiosaurs, all of whom (fossils tell us) became overtly 'eggless' in the era during which dinosaurs stalked the land. This begs an obvious question: why would so many water-dwelling amniotes evolve away from egg laying? What's so good about live birth in watery environments? One group of ocean-dwellers alive today offers some hints at why this mode of life may have been favoured in the waters of the Permian: the sharks.

Ancestrally, it is thought that sharks laid external eggs, like mesosaurs and others, but by the Permian and into the Age of Reptiles, this would gradually change in some lineages. Some shark species would evolve to keep their eggs in the oviduct (the 'egg duct') and provide them with very little by way of nutrients, flushing hatchlings from the reproductive passageway within minutes of their hatching there and becoming mobile. Other shark groups would evolve to offer more: in the uterus, they provided their young nutritious secretions absorbed by the umbilicus, probably much like the young of *Materpiscis*, the Mother Fish we met in the Devonian.*

Today, of 500 or so shark species described by scientists, about 70 per cent have evolved to give birth to live young, nourished either through yolk or through a placenta-like structure. The factors that have driven live birth in these species are comparable to those that would have influenced water-dwelling reptiles, including

---

* Other members of this group would continue to lay external eggs, which later specialised and have become known colloquially as 'mermaids' purses'. Made from thick, leathery parchment-like material, these eggs resemble pieces of ravioli. Modern-day rays continue to produce these egg structures, the largest produced by the aptly named big skate (*Beringraja binoculata*). At almost thirty centimetres in length, these egg packages stretch the 'purse' descriptive – you could almost fit a laptop in one.

mesosaurs of the Permian. First, when it comes to sharks and their embryos, size matters. Almost across the board, egg-laying sharks are smaller than live-bearing sharks and their embryos are far reduced. Larger, live-bearing sharks produce fewer offspring but they are far bigger than those of the egg-layers. Sand tiger sharks, for instance, when they are expelled live, umbilicus still attached, reach a length of almost a metre. At this size, they are already born as mid-level predators in most ocean food chains. Bluntly, it may be that the physics of the egg – how oxygen and carbon dioxide exchange upon its surface – cannot permit an embryo of such magnitude to exist outside of its mother. For a large apex predator, there is no 'choice' but to gestate the egg inside the body instead. This argument transfers well to the mesosaurs, ichthyosaurs and plesiosaurs, which were, on the whole, comparatively large.

Evolutionary biologists who study sharks also point to other potential benefits of live birth in some species: live-bearing sharks spend more energy on reproduction, sure, but, if environmental conditions change (for instance, food becomes scarce) these sharks can, in some cases, control the flow of nutrients into the oviduct, adapting their reproductive strategies to suit their changing world. Many live-bearing sharks can also reabsorb their embryos as a last resort. Egg-layers are not blessed with such an adaptable system: on the whole, they must fill their eggs with energy-rich yolk and expel them, no matter the changeable ecological conditions of their habitat. And so sharks, relatively more recently in evolutionary history, 'adjusted' their approaches to eggs just like we imagine mesosaurs, ichthyosaurs and plesiosaurs to have done in the past. The water can do predictable things to egg-laying vertebrates – it can draw the egg inwards. In time, partly for the same reasons, the land would do something similar to a small band of land-living amniote egg-layers that were the precursors to mammals. But we are getting ahead of ourselves.

The Permian landscape was still mostly a place for soft-shelled eggs, amniotic in form, laid by distant descendants of those earliest landfish, now evolving a more dynamic body shape for scouring forests and wetlands.

By the Permian, the ancestors of today's reptiles, birds and mammals had become far more than scuttering lizard-like creatures. Many were larger now and, as a group, these tetrapods (the four-legged descendants of Devonian landfish) were widespread and numerous. They had become a diverse group, split down the middle – some mammal-like, some more like the reptile groups that persist in modern times. These amniotic-egg-layers had come to shape their ecosystems in the same way that large mammals do today. By cropping the landscape, restricting the growth of trees and influencing populations of prey species these animals were, in their own way, building complex ecosystems. But this diversity would not last. For coming their way was a mass extinction event more catastrophic than any in Earth's history – the so-called Great Dying. In all, the incident, which took place approximately 252 million years ago, saw 81 per cent of marine species perish and 70 per cent of terrestrial vertebrate species lost.

The cause of the chaotic suite of extinctions at the end of the Permian Period continues to be discussed, but there is now general agreement among researchers that climate change, possibly linked to a series of sustained volcanic eruptions, baked the land

and poisoned the ocean through anoxia and acidification. The meteorite impact that occurred 66 million years ago, destroying the likes of *Tyrannosaurus* and company, pales in comparison to this creeping wave of death. This was a mass extinction so serious that even the insects were affected. The event saw eight or nine insect orders (the name given to large groups of taxonomic families) face extinction with ten more greatly reduced in diversity. Of course, being insects, the group bounced back: within 10 million years of the Great Dying they had re-evolved their former glory. Durable, waterproof, armoured – their eggs were, undoubtedly, one of the reasons for this.

Many of the insects that survived the Great Dying were noteworthy for possessing a new insect egg adaptation, one that evolved in the early days of the Permian and which the vast majority of insects use to this day. These eggs were different to those that came before, because the thing that hatched from out of them was not a miniature version of the adult form, like almost every egg previously, but something entirely more simple. At around the beginning of the Permian Period, an early life stage that resembled a grub hatched from an egg. It was an entity you could almost describe as a Permian maggot and it found new ways to extract energy from its environment before, later, metamorphosing into the adult, sexual, final form. This new life cycle – known as complete metamorphosis – took off spectacularly among insects during this Period.

In the modern day, insects that undergo complete metamorphosis (collectively known as the holometabolous insects) account for 80 per cent of all insects, equating to a staggering 60 per cent, approximately, of all animal species. So common is this life cycle, that it is almost certainly true that larvae (maggots, grubs and more) account for most of the animals alive on Planet Earth at this very moment. It is strange to think that, in this very instant, as you read these words, the majority of animal life is, apparently, in the waiting room, still working towards a final, egg-laying, sexual condition.

What were the first insect larvae like? And why did this curious strategy evolve in the first place? By looking at comparable features in the modern-day insects that undergo complete metamorphosis, we can be fairly sure that adult forms of this ancient insect had compound eyes and probably had sturdy, well-developed legs which allowed them to move upon the leaf-laden branches and twigs of trees which they likely consumed. It is thought that these ancestral insects resembled primitive earwigs, perhaps hiding under logs at night, keeping safe from scurrying lizard-like reptiles and predatory, dinner-plate-sized scorpions. It was here, with these insects, that the evolutionary transition to a middle state of insect life began.

Though worm-like, the earliest insect larvae had three distinctive pairs of legs at the front-most end and their mouths were equipped with chewing mouthparts. These mouthparts are particularly noteworthy, for larvae, in all insects that undergo complete metamorphosis today, use their mouthparts to gather nutrition. Importantly, insect larvae have dietary needs that tend to differ from those of the adult life stage. Those earliest grubs may have attacked fungus or rotting wood, but their adults may have sought tree sap or hunted other insects. This strategy, though subtle, unlocked an extraordinary ecological perk for insects. For the first time, immature and mature insects of the same species no longer competed for the same sources of food. Today, this is most eloquently expressed in caterpillars and butterflies: one life stage collects its nutrition from leaves; the other from flowers. This arrangement means that both life stages can live in the same place and prosper.

The most likely explanation for the evolution of complete metamorphosis is that it began through the simple mutation of a gene or two. This change saw eggs hatching earlier than they were 'supposed' to, before their development was completed, during a stage known as the 'pro-nymphal stage' in insects which undergo incomplete metamorphosis. It is during the pro-nymphal stage that the developing organism has legs and mouthparts, but it is long and thin, almost like a worm. One theory has it that for some early insects, pro-nymphs that prematurely hatched may have been able to use their mouthparts in the environment, feeding upon nearby substrates, perhaps even eggshell and egg yolk. If there was variation in the moment at which hatching occurred, which there surely was, natural selection could get to work on pro-nymphs, selecting those that made best use of their resources to continue their growth. And so, probably, the pro-nymph stage began to protract, bit by bit, in the insects that would go on to evolve complete metamorphosis.

There is good evidence to support this sequence of events. Particularly convincing is the fact that pro-nymphs and insect larvae have a lot in common. Neither possess armour, for instance. Neither has a thick, chitinous exoskeleton; instead, their bodies are soft. Inside, although their digestive systems are well developed, the nervous systems are simple semblances of the adult form. There are genetic switches common to both insect groups too. One, a gene named 'broad', is essential for the pupal stage – without it being switched on, caterpillars cannot become butterflies. But this gene is present in pro-nymphs too. In them, it controls the moulting process, how insects shed their old skins and grow new ones. Both 'classic' insects and those that undergo complete metamorphosis utilise the same hormones for growth as well. One, named juvenile hormone, moves pro-nymphs, as well as caterpillars, grubs and maggots, through their developmental stages.

The oldest known fossil of an insect larva supports this chain of events too. Dated to the start of the Permian Period (there or thereabouts) the fossil shows a nymph-like insect with broad, double-jointed mandibles, that also has features one might associate with larvae, including the familiar, distended, sausage-like body. This evidence, combined with that gathered from insect hormones and genes, suggests that the crossover to complete metamorphosis was another flashpoint moment occurring at some point at the interchange of the Carboniferous and Permian Periods. Today, all of the sturdiest evolutionary branches of the insect family tree arose from these early evolutionary trailblazers – the holometabolous insects.

Our world would look very different had this evolutionary innovation in insect eggs not occurred. Very basic ecosystem services that we take for granted – that dead animals disappear with time, that dead wood recedes, that dung doesn't hang around for long in the environment – occur so predictably because of the actions of insect larvae: maggots, grubs and the like. In fact, the evolution of the insect larval stage may have doubled the ecological complexity of forest food webs in the Permian landscapes. It may have been as important to the evolution of insects as the evolution of the serosa, refined and sharpened in the Carboniferous as it was through countless brushes with the elements.

But there is another plot twist for some of the insects that underwent complete metamorphosis back then. The fossil record suggests that, rather than laying eggs in soil or dead wood, some groups began depositing eggs both on, and later into, living plants. By adding special molecules to these eggs, chemical cocktails if you prefer, insects evolved that could modify the structures of plants, seeds or leaf buds, making them form 'nursery grounds' for their grub-like larvae. These structures are called galls.

Galls are hardened, mostly three-dimensional bobbles and bumps, resistant to most things, including many herbivores, that the environment might throw their way. Inside this protective shell, the young insect larva feeds, growing its way to maturity.

Many fossils of insect galls are known, particularly from the Permian. In some fossil specimens, the entry point for eggs can be seen; the exit holes from which prehistoric larvae escaped and metamorphosed. Bite marks and faecal pellets, scattered around like dust, can also be seen in these empty, fossilised egg chambers.*

First these prehistoric gall-making insects came for the ferns and grass-like horntails growing in those times; then, by the middle of the Permian, they evolved and came for the seed ferns and conifers. Hundreds of fossil galls are known, suggesting that this was a very

---

* In the modern day, it is worth remembering the economic and cultural value that galls have offered us. Oak galls are especially important in human history, for it was they, ground down in iron salt and gum arabic, that produced the earliest forms of non-soluble inks. For almost 700 years, right up to the nineteenth century, oak gall ink was the most commonly used type of ink in the Western world. As the entomologist Anne Sverdrup-Thygeson points out, we know the works of Shakespeare, Beethoven, Linnaeus and Galileo, thanks to the sacrifice of countless millions of pulverised galls, mashed to a pulp by pestle and mortar, their larvae often intact.

viable lifestyle for many insects of the time. As the Permian pro-
gressed, natural selection worked harder on the problem of how to
get eggs into plants. And then, in one or two insect groups, it went
one step further still, helping eggs get into other things, including
the bodies of other animals. For what we think is the first time, about
250 million years ago, parasitic insects evolved that could squirt
their eggs into animal hosts. In this group, a sharp egg-laying tube,
known as the ovipositor, became the refined tool for the job. These
insects became the ancestors of what we now call 'parasitoid' wasps.
Probably, at first at least, parasitism was a fringe activity for this
small group but, in time, the strategy would pay off.

The ovipositor, modified into a stinger in some wasp species,
is an egg-laying adaptation like no other, upgraded further in the
descendants of the early parasitoid insects – the wasps, bees, saw-
flies and ants. Shaped like a hollow needle, yet flexible enough to be
whipped from side to side, the organ allows for the precise delivery
of eggs into or onto hard-to-reach substrates. The 'wasp-waist' (the
narrow and flexible join between the thorax and abdomen) is actually
an adaptation – it allows the wasp to nimbly manoeuvre its ovipositor in a variety of directions. In many ways the ovipositor resembles
a surgical instrument.

Today, animals that these wasps insert their eggs into or upon
include maggots, grubs, pupae and adult life stages of many other
insects and, for some species, spiders. The parasitoid wasps that
target spiders are especially fiendish. Often, their grubs attach to the
external surface of the spider where they resemble an obese, immo-
bile blood-sucking leech. To stop the spider shaking off its intruder,
the parasitoid wasp injects an immobilising, paralysing nerve agent

into the spider after they lay their eggs. The paralysed spider is eaten alive by the grub. There are other villainous parasitic wasps in the modern day, including the (infamous) emerald cockroach wasp (*Ampulex compressa*). This wasp stings specific neurons in the brain of its cockroach host, switching off the parts of the host's brain that control the escape response, removing the cockroach's free will and turning it into a biddable slave for the wasp's grub to eat. The wasp walks the zombie, gently guiding it by its antennae, to the burrow in which it will lay its egg.

In modern-day wood wasps, the ovipositor is at the extreme edge of what natural selection can achieve. One record holder, *Megarhyssa macrurus*, has a 12-centimetre-long ovipositor that is more than twice the length of its body. There are even some parasitoid wasps that lay their eggs, Russian-doll-style, inside the eggs of other parasitoid wasps.

'I cannot persuade myself that a beneficent and omnipotent God would have designedly created parasitic wasps with the express intention of their feeding within the living bodies of Caterpillars,' Charles Darwin wrote in a letter to the American naturalist Asa Gray in 1860. More than a century and a half after writing these words, the point still stands. There may be no more accomplished, or more chilling, egg-layer in the history of our planet than the earliest parasitoid wasps finessing their style, some 250 million years ago.

The ancestors of wasps, small yet robust-looking flying insects with patterned wings and square jaws, litter Permian fossil beds from which have been extracted more than 15,000 fossil insect specimens overall. In slab after slab, one can observe silverfish, mayflies and griffinflies, some exaggerated to what might seem to us extraordinary size, including the dragonfly-like *Meganeuropsis permiana*, with its 70-centimetre wingspan – the largest insect known. Here, on these fossilised tableaus, insects that lacked a pupal stage – the 'classic' insects – mix with the insect upstarts, the grub-makers. The old-school open-jawed stoneflies, grinning mayflies and ancestors of dragonflies can clearly be seen, mixing with this new Permian in crowd, with their maggot-like larvae. In time, by way of species numbers, the in-crowd would continue to gather momentum.

Fossils like these provide a backdrop or a benchmark against which to compare the speed at which insects evolved into new species and how their extinction rates changed over time. Yet it is also through Permian insect fossil beds that we understand the severity of the mass extinction event that occurred in the final years of the Permian Period.

There is no doubting that insects were hit hard during this mass extinction, but being quick to reproduce, courtesy of their adaptive egg, this group recovered rapidly. Other animal groups had it far worse. The sea scorpions, whose ancestors may have, evolutionarily, seeded the land with arachnids, faced total annihilation. The trilobites, whose eggs had washed through ocean currents since the very earliest days of the Cambrian, were gone. Nearly every single species of sea anemone that lived, whose eggs, on moon-lit nights, formed a mist around coral reefs, disappeared forever. Through cnidarians like these, the last of the Ediacarans were almost totally lost to us.

After this evolutionary pinch-point, however, the world began to refill with the animals, and their eggs, that saw it through. It was the amniotes, the ancestors of reptiles and mammals, that fared noticeably well after the Great Dying, as well as, alongside them, another clutch of curious landfish descendants so far barely mentioned in our story. These animals, drawn back to water seasonally to lay their eggs, would appear, briefly, to take centre stage in wetlands once the dust settled on this period. In some of these strange land vertebrates, a different kind of metamorphosis was around the corner. Not maggots, but tadpoles . . . or something like them.

It was the turn of amphibians, with their jelly eggs, to make an unthinking play for the top spot on land.

Interlude
A post-Permian moment

In the arid, innermost expanses of Pangaea, among ripples of sand dunes hundreds of metres high, in the cracked mud of a once-thriving oasis, the hardest and most enduring of all animal eggs lay dormant. They rolled here in wind; they flew through dust storms that lasted for months to end up here. They are the only eggs, perhaps, for a thousand kilometres in each direction. But they exist in a strange state, somewhere on the continuum between life and death. The cells in each egg have paused. The cells of these bizarre sleeping eggs will only continue dividing once rain begins to fall and puddles begin to rise. At the end of the Permian, midway through an extinction crisis like no other, this happens more infrequently than ever.

Long ago, the rains came seasonally to this dry, desert core; then, later, the rains only fell in great storms every ten years or so. Now, at the height of the Great Dying, the land only floods in freak weather events, centuries or more apart.

Yet this lengthening gap does not matter to the embryos inside these eggs. A few days of rain, when it finally comes, a few tiny puddles or pools, is all each egg needs to un-pause itself and for cell division to continue. When this happens, from out of each tiny egg will come tiny fluttering shrimps with rows of leaf-like legs that flex in waves as they motor through the dank, salty, desert water. The entire life cycle of each shrimp will play out in days, in the time it takes each puddle to fill up and then evaporate. And then, when the puddle is gone, the next generation of eggs will rest in the sand, in the between-world, neither alive nor dead. The wind will pick these eggs up, just as it did the ancestors of those that laid them, and carry their lineage onward to the next century.

Through mass extinction events like the Great Dying, using this adaptation, these crustaceans will pass forwards. In time, they will travel through other extinction events, including the one that killed the largest dinosaurs. One day, they will surely outlast us.

It is a curious thought that the world in which we live could be an empty interlude – a blip devoid of thought or existence – in the before-life of an insignificant desert shrimp.

8

THE TRIASSIC TAKEOVER

Triassic Period, 252 million years
ago to 201.4 million years ago

'It may be hard for an egg to turn into a bird: it would be a jolly sight harder for it to learn to fly while remaining an egg.'

— C. S. Lewis (1898–1963)

Think of an amphibian egg in your mind's eye. Think of it in tapioca-like blobs that float on the surface of a pond. Picture the dots, one thousand or more in a single mass. Watch the low springtime sun, glancing off each glassy sphere; the faint wobble caused by the bite of a wintry wind still holding sway. Picture amphibian eggs glued in tidy masses to algae-covered rocks. Picture others laid singly, individually wrapped up one by one and glued to the tangling leaves of waterweeds. Amphibians – especially frogs, toads, salamanders, newts – lay eggs like these today. But this was an egg-laying tradition that was continuing to be refined in the early stages of the Triassic Period, 250 million years or so ago. For, after the Great Dying, the amphibians made a short-lived play for the land. They were one of three distant descendants of the landfish to do so in this period, the fortunes of which we'll unfurl here. By the time the Triassic ended, there would be only one winner.

Before we delve into the story of this triad, however, a reminder of the scale of the extinction crisis that these organisms had somehow survived. Moving up the strata in rocks dated from 255 million years to

250 million years, the extinction 'epidemic' is clear. More than half of all animal families were lost at this time, with as many as 95 per cent of all marine species. Seventy per cent of Earth's land animals dwindled to nothing. Fossil layers from this period finish up bleak; barren. The continents seem to have almost emptied themselves of all life.

The oxidised red rocks of the Triassic suggest that after the wave of extinction the animal recovery was slow. Mostly, these rock samples take the form of accumulations of sediments in western North America (Colorado, Utah, Wyoming and Arizona) and South America (particularly in Argentina, Colombia, Brazil, Uruguay, Paraguay and Venezuela) but they can also be found in the UK. I have spent many holidays, for instance, picking at the rusty sediments of Triassic outcrops on the south coast of England, not far from Torquay. Another hotspot is the Mendips – the limestone hills just south of Bath and Bristol in Somerset. Here, the rocks go from reds, associated with early Triassic desert lands, to the greenish-grey of mudstones laid down in water.

In fossils from these layers, the trials of egg-layers on land and in fresh water are exposed. Infrequent at first, then more and more common in the fossil record as the Triassic fauna begins to re-establish itself, one sees the stories of one particular egg-layer diversifying quickly to fill the empty niches of land. These were the early amphibians. Protected in their jelly casing and laid (we think) exclusively in ponds, lakes and rivers, these were among the first eggs to bounce back from the brink after the Great Dying.

Much of what we know about Triassic amphibians and the eggs they might have laid comes not from fossils of their eggs, but from fossils of these early animals sprawled out on sheets of brittle mudstones preserved in spectacular mating assemblages. Scenes, some scientists attest, of the moments right before egg laying occurred.

The first and most arresting of these fossil discoveries occurred in 1936. Like a bookmark poking out from a page, fragments of fossilised bone protruding from the rock face in a New Mexico quarry were noted down by husband-and-wife palaeontologists but not excavated. The find looked to be a jumble of skeletal impressions – some leg bones, vertebrae and a shiny, shield-like skull 1 metre long, pocked twice, one for each eye. The pair identified the bones as potentially rare amphibian fossils, notably large (up to 2 metres or more) in size. We now know these bones came from the extinct amphibian giant known as *Anaschisma*. It was two years before the pair could convince other researchers of the potential significance of the site, now known as the Lamy Amphibian Quarry. They returned with a team in 1938 and carefully, meticulously, unearthed a complete assemblage of *Anaschisma* with the help of picks and shovels and, in places, the measured use of dynamite. What was revealed was a wafer-thin fossil layer, just 10 centimetres deep, that ran for 15 metres across the exposure. Not just one individual, but a field of fossilised titans that belong to a group of ruling amphibians in the Triassic known as temnospondyls.

The zoologist Alfred Romer (of 'amniote egg' fame) was the scientist afforded the pleasure of first describing the fossils, which he wrote up as *An Amphibian Graveyard*. This is an apt description. In total, the death assemblage includes some 100 individuals, their round skulls each twice the size of a shovelhead; their eye sockets, like lost souls, stare out through time.

There are clues here about the circumstances that preceded this amphibian crime scene. All of the amphibians in this frozen moment are adults, all between 2 or 3 metres long, including the tail. Some 230 million years ago, these amphibians occupied the upper reaches of food chains, hunting much like crocodiles today, through stealth and the surprise application of powerful, fast-snapping jaws. Originally their deaths were deemed to be drought-related; that the individuals in the fossil array had gathered together at the last available watering hole in some long-lost shrinking Triassic oasis. But the fact there are no juveniles in this assemblage has led many scientists to propose that they were breeding individuals, perhaps preserved in the act of mating by an onrushing tide of brown, slushy flood water which dragged and re-ordered them on some muddy embankment further downstream.

What might their mating have been like? The fact that nearly all modern-day amphibians are external fertilisers – that males have no penis and instead release sperm onto eggs as the female releases them – suggests that these prehistoric giants probably did the same. But did they mate like newts and salamanders or like modern-day frogs? In newts and salamanders, males perform gentle, elegant mating 'court-ship routines' which often involve the male producing special 'love' molecules for the female, sent on their way, from a distance, by a rhythmic thrashing of the tail. In frogs, on the other hand, things are often far more raucous, with male frogs of many species evolving to grab hold of females from behind before spawning and rarely letting go, no matter how hard other males may try. (And they do try.)*

---

* Female frogs often use this to their advantage: if a female is grasped by a small male, in some species the female will simply carry the apparent 'weakling' to an encirclement of larger males, as if encouraging them to relieve her of her burden.

Currently, the fossil record remains silent about which strategy the giant Triassic amphibians may have incorporated into their mating. It may be, of course, that they evolved a mating style even more arresting and that some undiscovered fossil somewhere, currently eroding out of a rock on some weathered cliff face, is waiting to serve up the answer.

There are other fossil discoveries that hint at amphibian romances occurring in the Triassic. Some, like those known from Argana in Morocco (the Timezgadiouine Formation), show amphibian remains squashed into an ancient deposit, apparently also engaging in great revelry in some long-lost evaporating puddle. The fossilised remains of the amphibians in these assemblages show the largest individuals occupying the deepest middle regions of the shrinking pond. Outmuscled to the edges of the puddle, the smallest and weakest, presumably, were the first to perish in the drying Triassic sun.

Tadpoles (or rather, prehistoric amphibian larvae) are also known from the geological record, mostly from a group of large-bodied, crocodile-like amphibians, mentioned briefly before, known as temnospondyls. These fossils detail faint markings of vertebrae and limbs and, sometimes, show a whiff of tail. Some of the impressions are so fine that the remains of external gills can still be seen on the larvae, like faint feather boas; indeed, external gills are a physiological quirk that remains in the aquatic larvae of newts and salamanders today. These fossilised larvae do not particularly look like tadpoles of modern-day frogs; they are more like miniature, diminutive versions of the adult form. The metamorphosis that these temnospondyl amphibians went through was less dramatic than that we see in the tadpoles of frogs and toads today. The fossils show that the skull bones thickened slowly: the palette teeth receded gradually and the development of digits gathered pace late in development, as did the steady formation of bone from cartilage (a process known as ossification) as the adult phase approached. There is a scaly complexion to some of these palm-sized swimming larvae too.

After the Permian–Triassic extinction event, it was the crocodile-like temnospondyl amphibians that flourished and diversified most quickly across freshwater habitats. Many evolved to also take on terrestrial niches, stalking prey in wetland forests and shrublands, often overlapping with reptiles in their ecological roles. Some temnospondyls reached lengths of 4 metres or more. There was one fish-eating amphibian group, the trematosauroids, that evolved to live in the marine realm. Indeed, one amphibian giant even came to occupy the temperate forests of what is now Antarctica. But the amphibian uprising would not last long. By the middle years of the Triassic Period, competition was beginning to assert itself from other larger land vertebrates: the amphibians, it seems, would become constrained by the ecological dominance of the two other groups, both of whom laid amniotic eggs, still evolving since their first appearance in the Carboniferous: these were the dragon-like 'archosaurs' (ancestors of crocodiles, dinosaurs and birds, among others) and the more mammal-like 'therapsids' (which included among them the ancestors of all mammals today).

Perhaps surprisingly, after the Great Dying it was the ancestors of mammals that had, at first, the upper hand. These adaptable mammal-like reptiles (the word 'proto-mammals' is sometimes used by specialists) became, in many ways, cornerstone constituents of food webs and ecosystems by the middle of the Triassic. Likely, proto-mammals dug burrows for their eggs or laid them in tree trunks. Almost certainly, they were soft-shelled.

If we were able to look at these Triassic proto-mammals face to face, would we see anything of ourselves? We might see a simple covering of wispy scale-like hair on some parts of the body. And we might notice the beginnings of a familiar toothy grin common to nearly all mammals today: a mouth with incisors for snipping, canines for tearing, molars for chewing. In fact, teeth feature prominently in some proto-mammal groups. Among the most common in the Triassic Period were the cynodonts (meaning 'dog-tooth'), some of whom lived an omnivorous lifestyle comparable to that of modern-day badgers. There were also the dicynodonts ('two-dog-tooth'), tusked therapsids that were among the largest animals of their kind. Some of these barrel-shaped herbivores resembled wildebeest and were likely to have modified their landscapes in exactly the same way, by influencing the types of plants growing in each environment through their dietary choices. In fact, one single genus of proto-mammals probably accounted for most of the bony life forms walking upon the planet during the middle years of the Triassic.

There were also smaller Triassic proto-mammals too. Among them were insect-eating species that resembled, in a passing way, rats. The smaller proto-mammals possessed fur and almost certainly gave some form of milk-like substance to their offspring. Having evolved, through their fur, a primitive means to insulate their bodies from the cold, this initially diminutive group could explore, ecologically, a realm largely closed off to other land vertebrates: they became nocturnal. While, each night, the brains of their competitors cooled and their metabolisms slowed, these active amniotic-egg-layers seized new opportunities: they hunted insects in the dark; they crawled through leaf-litter, sniffing out millipedes and centipedes; they smashed snail shells against rocks, slurping at the soft insides. And they did this, largely, without fear of being eaten by reptilian day-walkers.

We take for granted how super-sensitive mammals are today: that their whiskers twitch at the smallest environmental changes; that their attentions can be re-shaped by altering gradients of wind-blown molecules pulled into their nostrils; that their eyes, wide in their sockets, can swivel like security cameras towards prey. These are night features, pure and simple, being crafted, out of sight, long ago, in the moonlit world. Probably, the rat-like proto-mammals were the earliest bony animals to walk the night.

Overall, back then, the proto-mammals (therapsids) were doing very well for themselves. In fact, if you were to have placed a bet on any group to see out the Triassic Period in pole position, it would surely have been the therapsids – our night-walking ancestors among them – that offered the best chance of return. Yet, spectacularly, it did not turn out this way. Because, as time tumbled forth and the Triassic entered its later phase, something rather unexpected seemed to happen . . . It was almost as if a hare-and-tortoise situation began to play out, with proto-mammals taking on the role (fittingly) of the hare and reptiles (naturally) the tortoise.

The fossil record details it clear as day: it was the archosaurs ('ruling reptiles') that would live up to their name and eventually diversify fastest. From the middle of the Triassic onwards, these reptiles were evolving forms that climbed and foraged among tree tops; forms with hard shells; long-necked fish-eating forms; marine forms that swam like sharks; herbivorous forms with hard, snipping beaks. Some archosaurs, in the form of pterosaurs, even took to the skies. And, among them all were the archosaurs known as dinosaurs, then still finding their way, evolutionarily speaking. Slowly, it seemed that amphibians and proto-mammals began to be edged out of their ecological niches. The reptiles were asserting themselves. The 'tortoise' was stepping up its pace.

Could eggs have been the reason for the ascendence of the archosaurs? Many of the traditional explanations for the success of archosaurs come down to features shared in modern-day archosaurs: the birds and crocodiles. If two distantly related animal groups living today have the same features, it is very possible that they inherited those features from the same ancestor long ago, and this can easily be proved using DNA.

One shared feature of modern-day birds and crocodiles is that, to save water, these animals can excrete urine as a paste (in their case, in the form of uric acid). Another once-shared feature is in the way that some modern archosaurs breathe: crocodiles are 'belly-breathers' – they possess interlocking 'pseudo ribs' across the belly that, when flexed, help pump extra oxygen around the body. (To evolve a more lightweight skeleton, birds have since lost this feature.)

The popular notion is that these two archosaur adaptations – urinating a paste and belly-breathing – helped early archosaurs see out periods when atmospheric oxygen levels dipped, such as in the end-Permian, or when arid conditions predominated and water became scarce. There are undoubtedly other factors too that might explain their success, not least the sturdier arrangement of limb bones that archosaurs evolved, but these two features are frequently noted in the literature, above others, as reasons behind the archosaurian advance.

Yet there is another feature that all modern archosaurs share today that some argue may also have been inherited from a common archosaur ancestor living in or just before the Triassic. This one related solely to eggs. All modern archosaurs, almost across the board, take good care of their eggs. Crocodiles and birds use vegetation to make their nests safe, just as we think dinosaurs and pterosaurs once did, to varying degrees. And, often archosaurs prepare these nests according to their environment, piling on leaves to insulate their eggs in cold periods or removing leaves to cool them when they become too warm – this is especially notable in modern-day crocodiles. In birds, temperature is managed through brooding – sitting on eggs to maintain the right conditions for the growth of the embryo. Many dinosaurs probably did this too.

Modern-day archosaurs – again, almost across the board – also communicate with their young, often *before* they even hatch from their eggs. These commonalities suggest that nest building and nest protection probably have their roots some time around the Triassic Period. It is hard not to imagine that this contributed in some way (at the very least) to the archosaurian advance – what palaeontologists like to call the 'Triassic takeover'.

Fossils of archosaur nests from late Triassic rocks of northeastern Italy's Dogna Valley allow us to visualise what a typical archosaur-dominated landscape may have been like.

Each morning, the sun, rising upon a rich mosaic of wetlands, would have bathed their nest sites in a fiery red. Archosaurs, stirred by the dawn, would have lain over or near their clutches of eggs.

Some would have looked like Komodo dragons or rather iguana-like; some looked like long-snouted crocodiles, some like crested, tailed tortoises. Each one would have greeted our rising star in the same way perhaps, horizontally aligning their body to absorb upon its surface as much heat energy as possible. Heat energy caused an upswing in metabolism for these animals. As their brains slowly activated, the parental urges of the archosaurs would have begun to take hold. And so, in a hundred different crater-like nests, some a metre or more in diameter, a hundred different snouts checked in with their embryos. Adults checked their temperature, the consistency of the substrate around them; they looked for evidence of nest predators. In some nests, on especially cool mornings, a scattering of new leaves and soil was likely to have been kicked into the crater-like construction to keep it warm. In some nests, the eggs would have been completely covered; in others, some of their surface would clearly be seen. The parchment-like covering of each egg would not have reflected light cleanly like the crystalline eggshell of a bird does today. Instead, the eggs would have been more leathery, damp . . . organic-seeming. As the sun continued to rise and the landscape became more animated, something akin to an archosaur chorus may have trickled through the air: a warring reprise of hisses, rumbles and bellows of competing archosaurs, each guarding, protecting, defending their embryos. Upon their nests, parent archosaurs would call and, beneath them, archosaur eggs would listen. There were likely to have been many, many days and many scenes like this. The archosaurian egg-layers would be around for a very long time.

Armed with adaptations and complex behaviours like these, the archosaurs became something of an unstoppable force in the latter stages of the Triassic. When the next cataclysmic event struck, which it did at the end of this period, it was archosaurs that came out of it fastest. And, most notable among them, were the archosaurs known as dinosaurs – a group that would diversify spectacularly in the Jurassic Period that followed. This extinction event was ecologically less trying than others in Earth's history and nowhere near as damaging as the Great Dying. But the event certainly led to a phase shift among those that laid amniotic eggs.

The trigger of the Triassic–Jurassic extinction crisis appears to be related to an extended period of tremendous volcanism occurring in the middle of Pangaea. In geological terms, break-ups are never easy. The breaking up of Pangaea into Gondwana (now Africa, South America, Australia, Antarctica, the Indian subcontinent and the Arabian Peninsula) and Laurasia (North America, Asia, Europe and Greenland), particularly, created a chasmic void that resembled a 10 million square kilometre lava field and that likely rumbled on for something like 40,000 years. In total, some 2 million cubic kilometres of lava may have been ejected into the atmosphere during this time, flooding the atmosphere with carbon dioxide that would affect the Earth's climate for hundreds of thousands of years. During this period, temperatures are unlikely to have risen in a consistent way. Instead, they probably see-sawed between extremes of hot and cold as rising sulphur concentrations (caused by volcanic eruptions) led to the formation of dense clouds which reflected the sun's rays back into space, cooling rather than heating the planet. In total, a quarter of all marine (animal) genera disappeared during this period and more than a third of all landfish descendants (a proportion of amphibians, archosaurs and proto-mammals among them) faced extinction.

The durability of each amniotic-egg-laying group was sorely tested during this time. Amphibians fared poorly, with the large crocodile-like temnospondyls knocked off their ecological perch, almost totally, forever. Those amphibians that survived the drama were the smaller, more frog- or salamander-like forms – the ancestors of the ones that share our ponds and pools today. Archosaurs declined slightly in the run-up to the crisis, but bounced back quickly, as we know. But the proto-mammals of this time are especially interesting. Of the proto-mammals to weather this crisis, it was mostly the nocturnal forms, ecologically straitjacketed by the presence of their archosaur competitors in the Triassic, that weathered the storm. And so, our ancestors were drawn closer, even more than they had been, to becoming creatures of the night. They evolved thicker pelts, larger, relatively more sensitive brains and eyes. And they evolved eggs that were protected, kept warm, within the body, for longer.

The Jurassic is where our own story begins. Back then, we were just one small clutch, isolated on one fragmenting continent, that knew moonlight more than it knew sun. It might never have happened were it not for the dominance of archosaurian nest-makers. It was through their ecological supremacy that our story would write itself. In time, natural selection would encourage the tortoise and the hare to race once more. And this time, famously, the result would play out quite differently.

9

NAVEL-GAZING

*Jurassic Period, 201.4 million years
ago to 145 million years ago*

'Who, my friend, can scale heaven?'

— The Epic of Gilgamesh

R ight before sunrise, when the insect chorus is intermittent and stalling, is when the egg thief is most active. It moves through the scrub lightly, deftly, barely casting a shadow on the ground. From low branch to branch, across twigs and around trunks, the urgent predator scampers; its whiskers twitch, its breath is fast and rhythmic. Its large eyes swing in their sockets, momentarily caught by movement above: a dusky silhouette on the upper reaches of a neighbouring cliff face, a nest . . . *prey*. Quickly the egg thief takes in as many details as it can. It sees a small pterosaur, crouched, high above, on a rocky outcrop, awaiting the warmth of dawn. The reptile sits upon a nest that is little more than a messy leaf-strewn platform pointing out to the sky. Faeces drip off the leaves and branches below the nest; a black goo rich in partially digested beetle wings. The egg thief has no interest in insects though. It pauses mere metres from the nest, its eyes firmly on the platform. The nest is poorly constructed; it is likely to be the pterosaur's first attempt. The female's colony-mates have taken the best spots around here, the tallest cliffs, safest from scurrying egg predators.

Faintly, through the cracks between twigs and leaves, the egg thief can see the pterosaur more clearly now – it is a small insect-eating *Anurognathus*, barely a match for the predator's size. The tiny pterosaur begins laying its eggs into its hastily arranged pile of leaves. In time, one, two, three, four soft, mucus-covered eggs fall like marbles onto the platform. The female pterosaur turns and faces each one as if to count them, sniffing and cajoling them so that they are wedged firmly into the detritus-heavy substrate. As the scent of eggs begins to waft down from the canopy, salivary glands in the approaching egg thief activate. Using its tongue, it smears digestive enzymes across the surface of each tooth. The egg thief moves closer; nearer, until it is within striking distance. It pauses, unmoving. The pterosaur mother looks tired; its flesh is stringy with no spare fat. Its small, compact torso, streaked with ligaments and bony tendons, shows little by way of flesh. This is a waiting game now; eventually, once warm enough, the young pterosaur will need to collect insects to stay alive, at which point its nest will be unguarded. By the time it returns to its nest, the eggs will be empty and the mucus-covered shells scattered.

For now, in this prehistoric crime scene, the identity of the assailant remains unknown. It is a mammal, certainly, one whose presence is given off in moments. In the crackle of leaves, a flash of tail, the trackways of foliage, trodden into mammalian highways between watering spots and warm burrows. But was it a mammal that laid eggs? One with a pouch? Or one with a placenta? Because early ancestors of all three types of mammal were around in the Jurassic Period and many, if not most, of them would have been partial to the eggs of archosaurs such as pterosaurs.

Although we think of the Jurassic Period as an Age of Reptiles or (more commonly) an Age of Dinosaurs this was, most notably, when the mammal group, descendants of the therapsids (proto-mammals), diversified their reproductive strategies.

The Jurassic was, for these newly evolving mammals, a kind of evolutionary playpark, where natural selection trialled new techniques for egg-making and rearing. Egg strategies became, almost . . . *flexible* . . . in this time. Mammalian evolution refused, for a time at least, to be put into a box.

What kind of world did these early mammals inherit after the turmoil of the end-Triassic? The early Jurassic Period saw the continuing separation of Laurasia to the north and Gondwana to the south. The world was warmer than it is today. Even close to the poles, forests grew. There were no ice caps and fewer arid regions. Ravaged by the end-Triassic extinction event, the fortunes of many reptile groups, some crocodile-like, some mammal-like, and some of the largest amphibians to ever have lived, had faded. One group of archosaurs, the dinosaurs, had adapted quickly to fill the ecological niches left vacant after this crisis.

Gigantic long-necked sauropods like *Diplodocus*, *Brachiosaurus* and *Apatosaurus* stomped through Jurassic cypress scrublands and coniferous forests, alongside *Stegosaurus* and its armoured, spiky allies. Even larger and more formidable predatory dinosaurs, some among the largest known, lived at the time. Our planet had never before supported land predators the size of *Allosaurus* – a stocky, bipedal predator almost 10 metres in length, for instance. Nimble dinosaur predators like the turkey-sized *Compsognathus* were becoming common too, as were the toothy, athletic ceratosaurs. By the

Jurassic, some dinosaur lineages were beginning to evolve harder eggshells, reinforcing their outer layers with calcite to guard them from egg predators of the age. But many, if not most, were still laying softer-shelled eggs. Crocodilians were pushed back into water, an environment to which, today, they remain wedded. Smaller amphibians flourished, insects too. And, in the shadows, the crooks, and the cavities of trees, small mammals prospered. The eggs they laid, many years ago, had once been amniotic eggs – the membranous egg crafted many millions of years before in the Carboniferous Period – but this form of egg was now changing, evolving down a trio of new evolutionary lines.

We, like most modern mammals, descended from the placental stock. For this group, embryos were kept protected in the womb, fed, watered and filtered of their waste through a new structure that was evolving at this time – the placenta.

For the first nine months of existence, you were in the cradling hand of this nurturing organ. Then, shortly after being born, you were torn apart. You survived the ordeal of birth while your placenta, the mammal body's only expendable organ, did not. Yours, the one that once nurtured you, was roughly 20 centimetres wide, 2 or 3 centimetres thick and weighed about half a kilogram. In structure, it had two surfaces. One, the basal plate, anchored itself into the womb lining and the other, the chorionic plate, faced you, the embryo. It is the chorionic plate to which your umbilical cord attached – a 50-centimetre lifeline through which the nutrients that made you once passed. If you could have looked between these two plates in the placenta, you would have seen a small cavity within which was an exercise in surface area maximisation. It was across thousands of numerous branching lobes, known as villous trees, that your de-oxygenated blood percolated from your mother and back, its oxygen supply restored. This wasn't all your placenta was for, of course. Far from it. One of the most important, and underappreciated, roles that the placenta undertakes is to be the pharmacist to the growing mammal embryo. In fact, the first thing the placenta does, in humans at least, is release hormones, specifically human chorionic gonadotropin (hCG), to stop menstruation from continuing. This is the hormone that 'pee-on-a-stick' pregnancy tests are activated by. In male embryos, these hormones are also what stimulate the testes to grow. Progesterone is another ingredient that the placenta provides. As well as helping the embryo implant into the wall of the uterus, this hormone prevents the uterus from contracting, securing a safe environment for the embryo's growth to continue. Oestrogen is also on hand as an extra hormone delivered by the placenta; its role is to increase the blood supply to the embryo in the later stages of growth.

How did this curious organ evolve? From where did it come? Watching the formation of mammal embryos in their very earliest moments (in humans, the first few days), when the cells are going from thirty-two to sixty-four in number, a layer of outer nurturing cells forms, known as the trophoblast. This is a feature common to all amniote eggs. In so-called placental mammals, however, something new evolved in these cells: as the embryo begins its first rounds of cell division, trophoblast cells start to interact with the cells in the uterus wall that are closest to them. These cells commence, in no uncertain terms, an invasion of the mother's tissue by growing finger-like projections that go on to digest maternal cells around them, encouraging maternal blood to leak out from spiral-shaped arteries beneath the uterus wall. Implantation, the often-used medical name for this process, downplays the drama somewhat – it looks ghoulish, parasitic, almost vampiric to the non-initiated.

Seemingly, in the Jurassic, mammals hit upon this strategy largely by chance: one key protein (known as syncytin), which the external cell walls of the placenta use to bind with maternal tissue, has its origins not in amniote egg-laying ancestors, but in something far more distant: a group of ancient retroviruses, of which HIV is now a part. It may well be that, at some point in the Jurassic Period or shortly before, an early mammal carrying a retrovirus in its egg cells accidentally 'unlocked' a more specialist means of maternal connection (or 'invasion' – choose your term) that began as a precursor to the mammalian placenta.

Today, placental mammals occupy top-tier real estate in ecosystems the world over, with one industrious placenta-endowed primate changing the entire face of the planet in mere centuries. For this reason, it is no surprise that the placenta has, in evolutionary terms, gained itself an exalted status. Yet, contrary to this narrative runs the fact that placenta-like structures occur commonly in many animal groups. You might remember the sharks from an earlier chapter, how the 'hungry' umbilicus, in some shark species, once starved of yolk, finds food from the uterus wall. And remember the rope-like umbilicus of the Mother Fish, *Materpiscis*, the armoured fish that once thrived in the Devonian Period. Live birth was particularly pronounced among the large, marine reptiles that lived in the Jurassic and it is likely that some (perhaps many) species evolved a placenta-like structure, the job of which was to provide nutrients for growth and remove waste from the embryo. There were other groups at this time 'toying', in an evolutionary way, with such an adaptation. It wasn't just mammals ploughing this evolutionary fur-row, but a variety of vertebrates on land and at sea. The Jurassic was, reproductively, an expressive period.

There is good evidence, for instance, that early scaly-skinned reptiles known as squamates (which include their descendants, the modern-day snakes and lizards) were also switching between egg-laying strategies in the Jurassic Period – some retaining eggs within the body like the early mammals and giving birth to live young, and some laying shelled eggs, like their cousins, the crocodiles. The commonality of live birth (viviparity) among distantly related snakes and lizards suggests that the condition has very deep roots indeed. In fact, undertaking DNA comparisons across species that live today, it is suggested that these reptiles 'clicked' into live birth (probably with a simple placenta-like structure in some groups) at various points in their evolutionary history and 'clicked' back out of it, according to the ecological conditions of the time. This switching in and out of different reproductive styles still happens in some lizards and snakes today, specifically when populations move between regions that are warm and cooler, more exposed, envir-onments. It seems that, on the whole, in cold climates live birth is an optimal strategy for modern reptile populations. For egg-layers, the selection pressures at work here are clear and obvious. Not only can cold temperatures slow the rate of cell division in egg-encased embryos, sometimes with lethal effects, studies have shown that eggs kept artificially cooler, even by a degree or two, produce offspring that are smaller at hatching, that grow more slowly and are less likely to survive their first year of life. The cold really does seem to matter for eggs.

And so, for many Jurassic amniotic-egg-layers, including early mammals, live birth may have been less about protecting young from predators and more about protecting young from extremes of temperature. The end-Triassic, with its see-sawing climate, may have been partly responsible, driving the gears of natural selection to churn out more live-bearing forms into the world.

Today, not all livebirthing reptiles have an organ analogous to the mammalian placenta, but many clearly do. In fact, this adaptation has occurred more than 100 times in the vertebrate line, probably many times unrecorded in the Jurassic. Evidence like this suggests that a placental bond between maternal and embryonic tissues is easier to evolve than we once imagined. In placental mammals, however, where the placenta is undertaking many jobs (feeding, cleaning, provisioning hormones), it is clear that the evolutionary ante was being raised in the Jurassic.

The vital connection that the placenta makes between adult and offspring in the mammal womb has added a peculiar twist to the story of the egg, with a fascinating evolutionary battle arising between mother and young. For it is in placental mammals, buoyed by the strength of this inter-uterine bridge, that the mother and embryo regularly try and renegotiate their evolutionary deals as development occurs. The mother's genes may 'desire' quality offspring, ideally many times over, but the embryo is out only for itself. This can lead to something of an uneasy truce between mother and embryo.

The problem occurs because both parties, young and old, are

physically connected for a long time, meaning that the embryo can, and does, evolve to 'beg' for more than it is being offered by mother as gestation continues. This 'begging' happens partly because of paternal genes. If embryos were clones of their mother, if there was no sex, no paternal genes, the evolutionary 'rules of play' would be simple: if embryos (in this scenario) evolved to take more at the expense of their mother, both players, mother and daughter, may suffer. In this situation, evolution would be drawn to a happy middle ground between the needs of mother and offspring. But the presence of paternal genes, with their own 'selfish desire' to spread, means that evolutionarily offspring may try to extract additional resources via the placental bridge.

The strange relationship described here was first put forward by Harvard's Robert Trivers in 1974, after he observed juvenile monkeys being weaned. The juvenile monkeys pleaded, they complained and found, in some cases, that it had been worth kicking up a fuss because the milk supply was continued for longer than usual: 'An offspring attempting from the very beginning to maximize its reproductive success would presumably want more investment than that parent is selected to give,' he wrote.

Trivers' idea, put another way, is that it is better for a baby to demand more of its mother by way of resources, even if that means the exhausted adult female may never breed again, than it is to take without complaining whatever is being offered. This phenomenon, known as parent–offspring conflict, goes some way to correcting the notion that offspring are passive recipients of parental investment. They are not. It wasn't long before other researchers looked at Trivers' parent–offspring model and began to consider whether the mammalian placenta might be a possible seat of influence in this negotiation between adult and offspring. It turned out that these researchers were correct. They discovered that the embryo does indeed often want more and identified the central role that the placenta plays to satisfy its goal.

From this research we now know, for instance, that in its earliest interactions with its mammalian mother, the embryo releases enzymes that burrow away further at maternal tissue – it already 'wants' more than is being offered. To stop the embryo in its tracks, the mother wields special compounds that inhibit, or nullify, the actions of the embryo's digesting enzymes. What results is, in the words of biologist David Haig, a system with a 'paternal foot on the accelerator and a maternal foot on the brake'.

This interplay between the wants and needs of paternal and maternal genes, occurring at the boundary between placenta and mother, explains why there is so much diversity between modern-day mammals in their styles and arrangement of placentae. Even in closely related mammal species, the placenta can differ in countless ways: it can be disc-like, leaf-shaped, ring-shaped, heart-shaped; its internal spaces can be labyrinthine or web-like, filled with ridges or finger-like branches. Each mammalian line is negotiating its own evolutionary contracts between adult and offspring.

Mammal embryos of different species vary widely in how 'invasive' their placenta have evolved to be. In some mammals, the placenta is very invasive, able to force itself all the way through the wall of the uterus deep into maternal tissue. In other mammals, it merely touches only the wall of the uterus, more like the placentae of sharks or some snakes and lizards.

What influences how combative a placenta becomes? Although an invasive placenta can help an embryo access more by way of nutrients, there can come a cost. Embryos with a more hostile placenta may absorb diseases carried in the mother's bloodstream, for instance, or they may be more likely to become victims of the mother's immune system.

By looking at how maternal genes activate in a range of mammals during pregnancy, and how these genes correlate with placental invasiveness between species, there is the suggestion that the ancestral (Jurassic) placental condition was also invasive. That, in the line that would become modern-day placental mammals (think: whales, rodents, monkeys, cats, dogs, dolphins, etc.), the placenta was fighting for all it could get. Was it something to do with their nocturnal niche? Or was it about bigger brains? Perhaps brains, evolving more prominent regions dedicated to smell and vision, required more energy to 'build' and this favoured the evolution of a more marauding placenta? The truth is hard to unpick, although it is an alluring question that we will return to in a later chapter.

This was not the only evolutionary innovation occurring in the Jurassic Period among the mammals. Because, somewhere around this time, natural selection turned out another, entirely novel, evolutionary routine. This clutch of mammals evolved mobile embryos, tiny grub-like young, which could journey, in their very earliest moments, from the mother's warm birth canal to the mother's warm pouch. They were the earliest marsupial mammals.

In the very earliest stage of their development, Jurassic marsupials connected with their embryos via a placenta in the uterus, but only for a very short period. These early marsupials started to give birth to smaller, far less developed, almost larvae-like, babies. They were embryos (or rather, neonates, for that is technically what they are) that could wriggle, climb, journey up the body towards a new safe place on the body, the pouch.

What defines this peculiar mammalian sub-group?

There is an image from my childhood that is burned into my brain. I can still replay parts of it now, so clear is the memory. It is a video of a red kangaroo joey being born, which was shown to me in a biology lesson. It was a short film known as *Birth of the Red Kangaroo* (1965) that some readers might remember. Within seconds of our teacher putting it on in the classroom, an uneasy hush fell over us all. Truly, we had never seen anything like it in our lives.

The documentary shows a thumb-sized joey, little more than a climbing jellybaby, somehow negotiating its way upwards across its mother's lower abdomen towards the lip of the pouch, where it clumsily crosses the threshold and clambers into the safety of its darkness. How does this pathetic neonate know where to go? How does a neonate with a brain smaller than a grain of rice navigate a slope so vertical? The classroom became flurried with both interest and silent horror in equal amounts.

Watching it now (it is freely available online), the film is more clinical than I remember it. There are no Attenborough-like dramatic pauses or anything very gentle by way of dialogue. It begins with footage of kangaroo smear tests, for instance, which I don't remember. And there is footage of semen being placed onto microscope slides for closer inspection. We see simple graphical representations and cutaways of marsupial female reproductive anatomy, which differs very much from our own placental system. To school children, this must have been especially confusing.

The details of marsupial reproductive anatomy are worth relaying for our story of the egg. As with placental mammals, marsupials have two ovarian 'horns' where mammal eggs are first delivered. Unlike in placental mammals, however, each marsupial egg passes through its own uterine space – marsupials have two wombs, in other words. This striking anatomical difference allows kangaroos and other marsupials to gestate two embryos of different ages at the same time. Indeed, most zoologists argue that it was probably the benefit of this arrangement that led to the evolution of the marsupial line in the first place – a manifestation of the proverbial 'don't put all your eggs in one basket' strategy. The two marsupial uteri connect to the same chamber, an open space known as the anterior vaginal sinus. From here, there are three apparent escape routes for any offspring being gestated there. The middle vagina (known as the median vagina) is the chute through which embryos travel downwards. The two other vaginas, which are located either side of the median vagina, are the channels through which sperm enters.

Birth of the Red Kangaroo goes on to explain what happens next, using real-time video of captive kangaroos. We see a pregnant mother-to-be kangaroo, frantically cleaning its pouch, an instinctual behaviour that occurs in female marsupials hours before birth. Then we see some shots of the birth itself, from another angle. Dazed, blind and apparently confused, the grub-like neonate reveals itself briefly within the female's reproductive opening before being obscured by the mother, who grooms the baby to free it from its amniotic shrink-wrap. Once clean, the tiny offspring springs into action, aligning itself with the pouch above it, orientated by scent. Then, armed with claws like grappling hooks, the baby makes its impossible ascent. As it moves upwards, the young marsupial, drooping its head from side to side, looks more like a newt or a land-walking fish than a mammal. The gravity acting on its bulbous head seems almost enough to see it tumble back down to where it originated. Yet, onwards and upwards it travels, clambering.

Finally, in the video, we see it enter the pouch and latch onto a teat, its thin skin revealing some semblance of organs or bones; no obvious hind legs, but definition in the forelegs and the familiar long, square jaws seen in all marsupial neonates. What an astonishing marathon with which to begin a new life. And it happens so early; some marsupial neonates make this journey just two days after gastrulation, when the cells of the embryo begin to differentiate into different cell tissues. At this stage, some marsupials weigh less than 5 grams, barely the weight of a sheet of paper. Comparatively, they resemble a human embryo at about eight weeks of development. Theirs is a most memorable adaptation indeed, first evolving, according to DNA comparisons of modern-day placental and marsupial mammals, at some point in the Jurassic Period.

When the tiny marsupial neonate makes its way into its new atmosphere, it has no eyes, although there is retinal (eye) pigment visible beneath its translucent skin and some semblance of earholes. The skin looks cellophane-thin at this stage, little more than a membrane to ensure that the tiny organism does not desiccate in the dry atmosphere of its new world. In a strange way, the tiny new organism looks like a perversion of both egg and animal at this point – an egg, with arms and a face, that climbs. The marsupial neonate even breathes like an egg, the bulk of its oxygen transferred directly through the skin. But look more closely at the neonate and you will see that some areas are clearly very well developed. The newborn marsupial forelimbs, for instance, are clearly toned and defined at this stage. Whereas the hind legs are simply buds, the forearms look strong and sturdy. This is to help it climb into the pouch. A reminder: no other vertebrate develops in this way. Another well-developed region is the jaws. Upon the face of a marsupial neonate can be seen the mouth or 'oral shield' that all newborn marsupials possess – a triangular hole that slots neatly onto the end of its mother's teat. The marsupial neonate is also born with a mature pharynx, larynx and tongue and there is even a well-developed epiglottis, presumably to prevent the newborn from choking on milk. From birth it is ready to climb and ready to drink, but that is all.

Marsupial milk is more than just food. It contains extra substances that placental mammal milk lacks, particularly novel immune system components. These include antibodies, neutrophils and macrophages, the white blood cells that kill and clean up bacterial and viral infections. Any damaging micro-organisms picked up on the journey between reproductive opening and the pouch are cleaned off by antibacterial secretions released by the mother in the pouch. Once attached, the mother's teat forms a bulbous swelling on the tip, which presses firmly against the inside of the oral shield within the joey's mouth, essentially locking its jaws into place. From this point onwards, for months on end, the joey's mouth, rather than its umbilical cord, becomes the entry point for the ingredients required for growth. In the Jurassic, this was a new approach – a kind of second placenta-like structure for the mammal group.

Traditionally, scientists assumed that marsupials were the 'primitive' mammal condition; that they somehow represented a 'half-way stage' in the evolution of placental mammals. More recent investigation, notably by looking at the arrangement of marsupial bones and how they develop in the embryo stage, suggest that it was the other way around – that marsupials are actually derived from very early placental mammals.

What benefit was this strategy? What advantage did jettisoning neonates into a pouch early in their development provide? Questions like these continue to nag at those that study marsupial evolution. The educated guess is that the marsupial 'way' is beneficial in landscapes, such as semi-arid deserts, where there is environmental uncertainty. In an unpredictable world, where resources may vanish without warning, the adult marsupial can easily abandon its tiny young and go on to fight another day. The placental mammal, on the other hand, is wedded to a more lengthy, internal gestation. There is also the fact that marsupials can nurse one offspring while gestating the next, a physiological double-hander that placental mammals cannot seem to match as effectively. It is possible that early marsupials filled up uncertain habitats with their babies more ruthlessly.

Either way, through innovations such as these, the Jurassic was a time of great mammalian expression and flexibility. Invasive and non-invasive mammal placentae; climbing mammalian grubs and . . . another style in this primordial playground? Yes, because there were the mammals that laid shelled eggs too. For monotremes, ancestors of modern-day egg-laying platypus and echidna, were denizens of Jurassic forests and wetlands as well.

Today, the monotremes remain a small band of evolutionary outliers. Just five species are known: the platypus, the short-beaked echidna *Tachyglossus aculeatus*, the eastern long-beaked echidna *Zaglossus bartoni*, the western long-beaked echidna *Zaglossus bruijni* and (discovered by western science in 1961 and named in 1998) Sir David's long-beaked echidna *Zaglossus attenboroughi*.

Monotremes differ from 'true' mammals because their embryos lack a placenta and females have one passage, rather than three, through which excretory products (including eggs) are flushed. This means that, like reptiles, monotremes (literally, 'single hole') lack an anus. They have no urethra and no vagina; instead females (like males) have a single hole – the cloaca (charmingly, 'sewer').

Just as with the proto-mammals from which monotremes evolved long ago, their eggs don't crack open like those of birds. They don't splinter into fragments. Instead, monotreme eggs are small and leathery, as if they are made of parchment. Platypus eggs are about the size of a large pea; echidna eggs, a marble. Unlike other modern-day egg-layers such as birds, the bulk of monotreme egg development takes place *in utero*, within the mother's body. Here, as well as yolk, the female produces nutritious fluids which pass through the egg's soft lining to nourish the offspring. Only after twenty-eight days of gestation may it finally lay spherical eggs that are kept warm in the burrow for another ten days before the eggs begin to stir and split open. Like lizards and crocodiles, monotremes possess an egg tooth to help them break free of their leathery shell. Once hatched, the young immediately begin the business of suckling. Lacking teats, monotremes produce milk directly from the skin, which hatchlings ('puggles') use to sustain their growth.

Monotremes differ from true mammals in other ways. Specifically, they have a shoulder girdle with bones not found in other mammals and their external ear openings are found at the base of the jawbone, as with reptiles, rather than at the top, as in the bulk of modern-day mammals. The plot thickens when their genes are brought into the mix. We know that monotremes possess the same identical gene (VTG2) that reptiles use to control yolk production, for instance, and that they also have a partial match for another (VTG1). We also know that many of the genes that monotremes depend upon to produce mammalian milk are found across true mammals and monotremes, confirming a shared common ancestor, most likely in or around the Jurassic Period.

It is staggering to believe that the monotremes have made it so long in a world dominated by other mammalian groups. It is like discovering an animal from the Lost World – like bumping into an evolutionary 'also-ran' from long ago. But, provided they can find a niche from which they cannot be displaced, animals like these can gain a knack for evolutionary persistence.

In a rather delicious evolutionary twist, the monotremes may have dodged extinction because of a quirk of their mammalian rivals' anatomy. Because newborn marsupials are required to climb and latch onto a teat very early in their development, it appears that the evolution of the marsupial body plan has been constrained. Big shoulders, clawed arms and bulbous embryonic heads make for poor swimmers as adults, it seems, hence why there never evolved marsupial whales (or, for that matter, marsupial bats). These marsupial forms could not evolutionarily 'progress', the thinking goes, because to evolve such a body plan would require the crawling embryo to change too dramatically. This anatomical confinement in marsupials was a good thing for water-dwelling monotremes, because it meant that they, rather than the marsupials, could evolve to take over waterside niches from marsupials, primarily in the form of platypus-like creatures (from which the echidna would later evolve). Hence, this small group of egg-laying mammals, free from competition, flourished alongside marsupials as they still do in parts of Australia today.

Three mammalian strategies; three different kinds of embryo. Whether they were egg-layers, pouch-wearers or true womb-bearers, the proto-mammals of the Triassic became, in the Jurassic, important entities in their own right. Only after the death of dinosaurs would they be inadvertently given the ecological freedom to do so more fully. But that was still a long time away, many millions of years, in fact.

For now, the Jurassic continued as an archosaurian egg-laying age. Large and small, the dinosaurs occupied the land; crocodile-like reptiles, some 2 or 3 metres in length, occupied fresh water; and pterosaurs – the flying reptiles – dominated the skies. Some pterosaurs were metres across; others, like *Anurognathus*, the pterosaur species with whom this chapter began, were small – not much larger than the largest of dragonflies.

In the later stages of the Jurassic, the nymph-like silhouettes of these tiny pterosaurs were very common indeed. Throughout the day, while the sun shone bright, their shadows sped like arrows in dappled lines across the forest floor. Yet by evening, when the shadows of the trees and shrubs grew long, their activities subsided. The sun went down and pterosaur brains, starved of heat energy, became sluggish and sleepy. It was at this point that mammal burrows, many metres below, began to stir with activity.

In one cosy burrow in the cleft of a tree stump, one by one, naked proto-pups are expelled from their mother. In their leaf bed, the newborn mammals squirm and roll over one another, in uncomfortable, rapturous anticipation. Ravenously, the pups focus on a single task: to feed. The fats that leak into the mouths of these babies from the naked underside of their mother contain nutrients reconstituted from the mother's meal earlier that day. The archosaur eggs, stolen from a pterosaur's nest, now provide a rich soup upon which new life is being nourished. But soon, the mother will be empty. So while the pterosaurs sleep, just before sunrise when the insect chorus rises and falls, the egg thief will return.

10

THE ART IN THE ISTHMUS

*Cretaceous Period, 145 million years
ago to 66 million years ago*

A brand, or flaming breath
Comes to destroy
All those antinomies
Of day and night

– W. B. Yeats (1865–1939)

The Cretaceous was the era in which the Earth's continents would have become recognisable to us today. The great land masses of Africa, Asia and Antarctica, still connected to Australia, had by then assumed a decipherable character, yet there was no snow or ice at the poles. North and South America, now bridged, were still unwedded back then. Rising sea levels saw North America gifted its own expansive inland ocean – the Western Interior Seaway, a body of water some 3,000 kilometres long and, in places, half a kilometre deep. This ocean ran like a scar across the continent, flanked to its western side by what is now the Rocky Mountains. The continent we think of as Europe was, back then, mostly unrecognisable, consumed as it was by a warm, inland sea. Parts of what are now Spain, France and Italy poked through its surface, cupped to the north by the mountains of Scandinavia, slowly eroding into fine muds and sands. Europe was a continent of islands back then. And, it is upon one of these islands that the story of this chapter begins.

Our perspective shifts now to a noisy bustling colony of Cretaceous birds. These birds are known as Enantiornithes – a group sometimes called 'opposite birds' due to the topsy-turvy arrangement of the shoulder bones, which contrast sharply with those of modern birds. The opposite birds are the most abundant birds of the Cretaceous. And – aside from their teeth and their clawed wings – these ones resemble gulls. Each year, Cretaceous opposite birds like these fly over cycads and seed ferns and conifer forests to get to this island. They know the outline of the ocean; the make-up of each rocky face; the shape of each coast. They see the ripples and folds of the land and, eventually, the entry point to their nesting grounds: a long and dusty sandbar behind which there is an expanse of baked grey mud where their kind have nested for generations. On a single nest in that colony sits a single bird, bringing into existence a single egg.

This egg, remember, is an amniotic egg – a Carboniferous innovation that 'land-proofed' the egg. By way of a reminder, the amniotic egg is a series of fluid-filled membranes that includes the amnion (which protects the embryo), the allantois (assisting with waste disposal and gas diffusion), the yolk sac (energy source) and the chorion (an overall enclosure for these structures). Together, in birds, these structures float within a jelly-like layer, the albumen, which is itself enclosed by the outer shell layer. As we have seen in previous periods, the shell layer is, in most reptiles, soft, brittle and parchment-like. Not so in birds. Their eggs have hard, calcified shells; shells so tough that, occasionally, they could even survive the process of fossilisation. Anatomically, the story of how these hard eggshells are made begins deep inside the female bird's body.

We start in darkness, right at the beginning of the female's egg tube, the oviduct. Here, the fertilised egg sits, soft and vulnerable; still awaiting its chassis. The spaces here in the oviduct are tight. To cajole the egg forwards, rhythmic waves and contractions travel in ripples down the tube. Through this activity, the fertilised egg is encouraged into the maze that will make it.

If you could look at the egg at this point, so early in its journey, you would be struck by the obvious yolk, surrounded by a thin layer of watery fluid, kept within a mostly water-tight bag, contained within a soft membrane. The viscous fluid, known as albumen, is especially important in the formation of the egg. For it is here that the egg derives its safety features. The albumen gives the embryo and yolk an airbag-like protection as it is squeezed down the egg tube, but it will also provide protection to the embryo when the egg is laid. Albumen is rich in an antibacterial molecule known as lysozyme, along with more than one hundred other proteins whose job it is to protect the embryo and its yolk from bacteria and other microbial agents.* Protected by the albumen, the egg is nudged down the tube to its first stop – the isthmus.

Before the egg is provided with its crystalline chassis, it requires a foundation layer upon which the shell can be added. This foundation layer is added to the egg by glands located here. As it passes through the isthmus, the glands spray the egg from all sides from

---

* In *The Most Perfect Thing*, Tim Birkhead's arresting and authoritative book on bird eggs, to which this chapter owes a debt, the albumen offers a kind of purposeful 'nothingness' to the embryo: 'To a microbe, the journey across the albumen from the shell membrane to the yolk on the inside is equivalent to a human trying to walk across the Atacama Desert.'

tiny pores with fibrous strands of proteins, including collagen. The external surface of the egg becomes a profusion of busy threads and fibres crosslinked, meshed untidily, but with purpose.* The shell layers are added in the uterus of the bird and, again, are applied by tiny pores in the lining of the tube, linked to sacs of chemicals that erupt or burst when compressed by the passing of the egg. The first of the chemical treatments added here sees the release of a chalky solution. This calcium carbonate soup looks, at first, foamy. It splurges onto the balloon-like surface of the egg, forming soft, doughy pillars, which immediately begin to harden. Next come more aerosols – mostly water. These molecules find their way between the pillar-like structures and travel through the egg's membrane towards the albumen. As another dose of concentrated calcium carbonate is added, the egg swells.

Hour after hour, these sprays are delivered in protracted bursts, creating a network of columns stacked closely against one another. Tiny vertical spaces remain in the structure of the shell at this point, and it is these that become the pore canals – the avenues through which gases and water vapour can move back and forth after the egg is laid. This calcium-rich chemical treatment is, almost certainly, one of the egg's most costly ingredients. To find enough calcium to produce this layer, many female birds gain an insatiable interest in items they would never normally consider food. Potential items include fish bones, the sloughed exoskeletons of crustaceans or fragments of mollusc shells. This hunger for calcium can drive birds to trial other novel dietary additions. In the modern day, small birds occasionally

---

* In a boiled egg, this is the paper-like layer that sits just under the shell.

pick through the faeces of predators, for instance, swallowing any fragments of bone they can find. Snails, whose shells are particularly rich in calcium, are another favourite. Eating mortar from brick walls is not uncommon. Sometimes, in desperation, birds swallow sand. Calcium foraging in birds seems to occur in the evening, rather than the day. This is probably (according to Tim Birkhead in *The Most Perfect Thing*) so that calcium can be assimilated into special calcium-storing medullary bones in the night, ready to be drawn upon the following day.

Medullary bones occur in dinosaurs too. And for good reason. Because (as any six-year-old dinosaur obsessive will tell you) birds *are* dinosaurs. Or rather, birds are an offshoot from the dinosaur family tree, specifically from the dinosaur group known as thero-pods – broadly, the two-legged tribe that includes *Tyrannosaurus*, *Velociraptor* and company. It is with this dinosaur group that hard, crystallised eggshells became commonplace. This calls for a brief digression from our story of the development of the hard egg of the female opposite bird. We need to move, momentarily, onto dinosaurs.

That dinosaurs laid eggs at all was, for decades, a dubious call to make. William Diller Matthew, curator of the American Museum of Natural History for thirty years from the mid-1890s, was fond of speculating that dinosaurs partook in live birth, in a mammalian fashion, carrying offspring to term, one at a time, like elephants do today. It was in the 1920s that the first evidence of dinosaur eggs was found, on a research expedition in Mongolia that sought to find fossil evidence of the earliest humans. The expedition lead, Roy Chapman Andrews, recalled in his field notes that:

On July 13, George Olsen reported at tiffin [lunch] that he had found some fossil eggs . . . We felt quite certain that his so-called eggs would prove to be . . . geological phenomena. Nevertheless, we were all curious enough to go with him to inspect his find. We saw a small sandstone ledge, beside which were lying three eggs partly broken . . . The brown striated shell was so egg-like that there could be no mistake. Granger finally said, 'No dinosaur eggs have ever been found, but the reptile probably did lay eggs. These must be dinosaur eggs. They can't be anything else.' The prospect was thrilling, but we would not let ourselves think of it too seriously, and continued to criticize the supposition from every possible standpoint. But finally we had to admit that 'eggs are eggs,' and that we could make them out to be nothing else. It was evident that dinosaurs did lay eggs and that we had discovered the first specimens known to science.

Expeditions in the Gobi Desert have since recovered many more nests, frozen in time, their unhatched inhabitants victims of flash floods during the late Cretaceous Period. These nests contain sometimes fifteen or so eggs, apparently laid in pairs. From discoveries like these we see familiar patterns among their styles: some are spherical in shape, others are elongated. Some have latitudinal symmetry, others are pointed at one end like a chicken's egg. Elongated eggs appear to be especially common among one dinosaur lineage, the theropods. But there are important distinctions between the eggs of (non-bird) dinosaurs and birds. Dinosaur eggs differ from those of modern birds because they are often ornamented with strange

patterns. Networks of nodes and ridges occur across the outer surface of their fossil eggs, making some look pitted or even cratered.
Some of these calcified vessels are covered in net-like patterns, which
may have added strength to the egg. In some examples, these patterns gather at the poles of each egg, leading researchers to speculate
that they may have served a role in expressing carbon dioxide from
the embryo. Some dinosaur eggs have longitudinal lines, straight
as a ruler; others have wavy lines that are almost rippled like the
surface of a pond. The function of these mysterious ornamentations
is still discussed.

Either way, the calcified shells of many dinosaurs and birds meant
that springtime was probably an urgent season in the Cretaceous,
with many individuals and species developing a short-lived obses-
sion for the elusive ingredient, calcium. Even dinosaurs we think of
as herbivorous may have been driven to gnawing on carcasses or
chewing on mouthfuls of washed-up oyster shells during this period.
Perhaps some large dinosaurs may have been partial to eating the
nests of smaller dinosaurs? It's certainly possible; likely, even. But,
for now, our digression into other dinosaurs over, it is time to return
to the dark, suffocating tube where our Cretaceous opposite bird egg
is being constructed. What is happening there happened in dinosaurs
too. In fact, if you are reading this in spring, it is occurring right now
in the birds with whom you share the neighbourhood.

Once the calcium is digested and reassimilated into glands in
the uterus of the egg duct and liberally applied to the egg, the shell
hardens into a cement-like armour covering. This is now a shelled
egg, in the pitch black of an egg tube, slowly getting closer to meet-
ing its world.

Yet our egg is still not finished, for there is paintwork yet to be applied. For this, the egg travels down the latter stages of its factory line towards another region of the egg tube, one laden with even more nozzles. The chemical constituents sprayed in this lower part of the tract add aesthetic rather than structural detail to the egg. It is here where coloured dyes are added to the eggshell, which mix with the calcium carbonate layer to create the egg's occasionally rich and vibrant colours.

The egg, now decorated, is gifted its final flourish in the last stages of the duct. Here, a chemical treatment of glycoproteins, polysaccharides and lipids is delivered, creating a waxy layer, the job of which is to deter and kill the microbial invaders it is destined, within moments, to meet in the outside world.

The delivery of the egg is swift. It passes through the cloaca of the female opposite bird with a gentle thud and falls into the mud. A toothed beak-like jaw gently nudges the egg firmly. Parental eyes assess its condition, nostrils confirm its chemical signature and a brain remembers the exact patterns and arrangements of colour on the egg. The egg is *identified*; seen, for the first time, by the bird that laid it.

It was during the Cretaceous Period that, for the first time in Earth's history, eggs began to communicate visual information to mother, father or both. Indeed, this is a feature that remains in many birds that live today. One sees it in the glossy blue of a blackbird egg; the rosy red eggs of the Cetti's warbler; the Jackson Pollock designs of the rock bunting; the malachite eggs of the emu; the tea-stained splotches of the mockingbird or the chaotic pencil-line etchings of the yellowhammer. From the Impressionist simplicity of the dunlin's egg to the cloud-like markings on the eggs of the sandwich tern, it is clear that eggs have something to say. These colours and markings on eggs were once considered a uniquely bird thing; we now know that this is, more generally, an adaptation pioneered by dinosaurs, particularly the bipedal dinosaurs known as theropods, such as *Tyrannosaurus* and *Velociraptor*, from which birds evolved.

The first discovery of dinosaur egg colours was in 2015, through research undertaken by molecular palaeontologist Jasmina Wiemann of Yale University. Wiemann and colleagues looked at the fossilised eggs of oviraptor and sampled them for trace biological compounds associated with colour in modern bird eggs. What the team found was two important compounds in oviraptor eggs: protoporphyrin and biliverdin. In birds, these compounds tally with eggs that are blue-green in colour so, the team concluded, oviraptor eggs were probably the same. Does this mean that all dinosaurs like oviraptor had colourful eggs? To answer this question, Wiemann worked with Mark Norell, palaeontologist at the American Museum of Natural History, to devise a non-destructive technique that allowed the constituents of fossilised dinosaur eggshells to be examined. The technique they came up with, known as high-resolution Raman microspectroscopy, sees computers analyse, through lasers, how light interacts with the chemical bonds of specific molecules on the surface of the eggshell. Using this technique, the research team investigated fossilised eggshells from fourteen other dinosaur species. What they found was that theropod lineages, including Cretaceous birds, had colourful eggshells, but those of other, more distantly related groups – *Brachiosaurus*, for instance – did not. *Velociraptor*, a theropod dinosaur that aligns closely with the theropods that evolved into birds, laid painted eggs that were speckled, dusted with spots like those of a sparrow.

All theropod dinosaur eggs (including birds) are coloured by varying the concentrations of the two chemicals identified by Wiemann – protoporphyrin and biliverdin. Protoporphyrin (or more specifically the protoporphyrin variant known as porphrin IX) provides the reddish-brown colour and biliverdin, a breakdown product of haemoglobin, the greeny blue. The application of these chemicals, courtesy of pores in the lining of the uterus, differ between species and, sometimes, individual birds. Banks of pores produce splodges and stains; single pores produce lines or speckles. Blotches, colour bleeds and nebulae-like clusters are common on dinosaur eggs, including those of birds.

Yet what might eggs be 'indicating' or 'saying' through markings such as these?

For decades, Darwinists wrestled with this question (then applied only to modern-day birds) and got nowhere. Alfred Russel Wallace, co-discoverer of natural selection, believed that eggs most likely evolved patterns and colour to camouflage themselves and, in some cases, this is undoubtedly the case. The eggs of little terns and ringed plovers, left in scrapes on pebble-covered beaches are, no doubt, adapted to their environment. In fact, once or twice, I have almost trodden on their simple nests, so well blended are they into their surroundings. To support his hypothesis that colour patterns were mostly about camouflage, Wallace observed that eggs kept in dark places, such as the eggs in a kingfisher burrow or a woodpecker hole, tend to lack colour or markings, a pattern that does, indeed, hold true across many bird species. But then what of brightly coloured eggs, like those laid by the American robin, the blackbird or the song thrush? The colour of these eggs makes them positively yell out to predators from the nest. Could there be something else going on?

Perhaps egg colour is to do with sexual selection, others argued. Interestingly, there is evidence to suggest that males of the American robin tend more readily to chicks that hatch from the brightest eggs, which could indicate that females invest in brightly coloured eggs to assure the male of its genetic fitness, perhaps to quell any ideas the male may have of (genetically) feathering other nests. So, there is evidence that egg colours could be about camouflage and, in at least one species, that these colours could be advertisements of mate quality. But there is clearly more to the puzzle than this. In 2014, a multi-national team of researchers published evidence to support another idea: that bird egg colours give embryos protection from the sun. This hypothesis has it that, through eggshell pigments, eggs can be afforded a form of natural sunscreen, protecting the growing embryo from the sun's radiation. Using egg samples gathered from seventy-four bird species and investigating, using a spectrophotometer, the amount of photons absorbed as light passes through the shell, their results clearly showed that eggs laid in locations that received a lot of sunlight had more pigment than those eggs tucked away in burrows or holes in trees.

As interesting as these three hypotheses are, the problem is that there are many exceptions to the apparent rules. Ostriches, for instance, lay pure white eggs, yet they lay them in a nest that is baked in desert sun for much of the day. Where is their sunscreen colouring? Or, for that matter, their camouflage?

Another hypothesis has gained support in recent years, however. This one argues that brood parasites (nest cheats, for want of a better phrase) are what cause egg colours and patterns to evolve. In this argument, the egg covers itself in signature patterns so that parents are able to work out which eggs belong to them and which eggs may belong to those of 'cheats' – more specifically, nest (or brood) parasites, such as cuckoos. We have, as yet, no idea what these Cretaceous egg cheats looked like. It may have been that there were a handful of 'cuckoo' species or many.

The word 'cheat', often used by evolutionary biologists, requires some explaining. The fact that cheating is, evolutionarily, a phenomenon seen across animals, big or small brained, proves that cheating is not about forethought or considered planning. Some mite species, for instance, change their cuckoo-like behaviour depending on the threats to their offspring. If egg predators are not around, they breed freely but, if they should appear, the mites change their strategy, laying their eggs in the egg clutches of mite species more adept at protecting their young. Animal cheats are not 'cruel', either. Many cheats probably begin life as stressed, desperate individuals, out of options that then become genetically 'liberated' once the strategy proves successful and establishes itself. The benefits to this way of life, of becoming a brood parasite, are obvious. Brood parasites no longer need to invest as much in rearing their young as other birds. They are free to make more eggs and better guarantee their genetic perpetuity in so doing.

Well-known examples of brood parasites include the indigo-birds, the whydahs, the cowbirds of the Americas and, as mentioned, the cuckoos of the Old World. Some of these brood parasites lay their eggs in the nests of others of their own kind; some evolve to lay their eggs in the nests of other bird species entirely. Both strategies see brood parasites evolve the same adaptations. Birds that lay their eggs in other nests often evolve thicker eggshells, for instance, perhaps to protect the parasitic embryo from suspicious foster parents, seeking to puncture the shell surface or to crush the eggs when sat upon. These birds also tend to evolve offspring that hatch more quickly than their hosts' eggs. If the parasitic chick is fed first, long before their 'siblings' have even hatched, the parasite stands a better chance of outmuscling its nestmates. This stage can be brutal for the host's offspring. In the common cuckoo (*Cuculus canorus*), for instance, the young cuckoo physically kicks the hatchlings of the host bird out of its nest when left alone by the parent birds.

Often, brood parasites evolve eggs that look like those of their host species (which is little surprise) but, of course, this opens up the door to an evolutionary counterstrategy. If the host species can keep changing the colour of its eggs, generation by generation, it can better spot the eggs of intruders, who may always be one step behind. Thus, across many species, an 'arms race' can develop between host and parasite: every advancement in host egg 'design' forces the brood parasite to adapt its mimicry and every advancement in mimicry forces the host to hit upon a fresher 'design'. This means that, without any opinion or intelligence or any artistic nous of any sort, natural selection can bring into being something so beautiful as a patterned egg. Eggs like these are of no use unless they can be understood by parent birds, of course. And so, with the evolution of brood parasitism, birds (and probably many theropod dinosaurs) evolved systems to spot eggs that are not their own. Quite possibly, they became just a little bit smarter.

But that's not all that is going on here. An interesting addition to this magnificent evolutionary dance is that some host species may recognise that they are harbouring a parasite in their nest and yet continue to rear the intruder regardless. There is evidence, for instance, that the great spotted cuckoo (*Clamator glandarius*), whose hatchlings do not kill other chicks in the nest when they hatch, periodically revisits the host's nest to check that the parasitic chick is still alive. If it is not, the great spotted cuckoo attacks the host nest, killing all of the host's offspring. By partaking in this strategy (called, for obvious reasons, the 'mafia hypothesis'), raising one parasitic hatchling in each batch of eggs is a better return than risking the murderous wrath of an angry cuckoo. In some cases, there may even be benefits to having a parasite chick around. Great spotted cuckoo hatchlings, for instance, secrete repellent odours when nest predators approach. These, when they rub off on the host's offspring, are thought to protect the nest as well, increasing rather than decreasing the survival chances of all of the nestlings.

It is highly likely that in the busy sea-facing shoreline communities of Cretaceous birds these different forms of parasitism were being trialled and tweaked for the first time, with novel egg patterns and colours. Way back then, cheats were already trying to stack their luck.

Retrieved from the Gobi Desert, a single fossil slab offers a hint as to what such cheats may have looked like. In the slab are the remains of a bird-like oviraptor draped over its nest in a protective embrace. And below this, among its hatchlings, are the partial skulls of two small dromaeosaurids – sickle-clawed dinosaurs that resembled *Velociraptor*. Were they the remains of prey . . . or the hatchlings of a brood parasite? More fossils like this, discovered from the Gobi Desert or elsewhere, will surely shed light upon this fascinating phenomenon.

In my opinion, eggs have been too often overlooked in popular interpretations of the Mesozoic Era, the so-called Age of Dinosaurs that consists of the Triassic, Jurassic and Cretaceous Periods. Dioramas and thematic Hollywood set pieces have giant dinosaurs wrestling one another or silhouetted on high things roaring into the sky for nothing more than dramatic effect, yet (apart from in a handful of cases) their eggs rarely feature. The reality is that a stroll across a Cretaceous floodplain would, depending on the season, likely be a theatre of egg-related dramas: the desperate search for calcium on some moonlit evening; the volley of mating rituals and calls; the scraping of nests, the digging of burrows, the covering and patting-down of soft muds and sands; egg camouflage, egg parasites; counter-strategies and evolutionary arms races. We have no reason to suspect it could be any other way. This is, after all, what we see with birds, those familiar dinosaur descendants, each spring.

We think of the dawn chorus as a thing of beauty and wonder, but to birds it is a chaotic, violent, desperate seasonal urge. One only needs to look at how ragged and worn out many birds look when they raid birdfeeders in our gardens in these months. I am often struck by how tired and tatty blue tits look by late spring, when their demanding chicks, hungry for caterpillars, seem to have almost stripped adults of their life force. Perhaps *Tyrannosaurus rex*, tending to its offspring, was the same? Not roaring with anger, just . . . well, very tired indeed.

Protected from microbes, armoured against the environment and decorated to be remembered, the theropod egg became a very complex egg indeed in the Cretaceous. But the tale of the egg which began this chapter is not complete, for it has not yet hatched. What begins deep inside the oviduct of the female bird, where its shell layers are gathered and upon which its character is painted, culminates in this, the egg's final act – the hatching.

Hatching is perhaps the most coordinated and practised moment in the history of most amniotes. In birds, trapped in their crystalline cases, it is an action that has been refined a little compared to other species, but the basics remain. The defining moment for the chick actually happens in the period just before it hatches from the egg. At this moment, its limbs collected underneath its body and its beak tucked underneath its right wing, the pre-hatchling begins to squirm. The network of blood vessels that line the inner surface of the shell, that grasp at the oxygen entering the egg through its pores, begins to shut down. The blood in this network is reabsorbed into the body of the embryo, which starts to seal itself off at the umbilicus. The developed chick severs its connection with the yolk at this point, draining what is left of the yolk sac directly into its small intestine. The chick readies itself for its first breath now. To take it, the agitating chick moves its head out from underneath its wing and penetrates the hollow chamber at the end of the egg with its beak. This is the air space at the tip of the egg, visible when removing the shell of a boiled egg. It is here, before the chick hatches, that it fills its lungs for the first time. This breath is the single most important one in the life of a bird. Without it, the chick has little chance of breaking free from the egg; it energises and empowers the young life for what comes next.

Now, the chick pushes its beak hard against the internal surface of the roof of the egg. To break through, the chick is aided by the 'egg tooth', the calcified structure common to many amniote egg-layers. It pecks, once, twice; again, harder. Finally, a crack appears in the shell, small at first and then, with a few more precision pecks, it becomes larger. Through this hole, a great gust floods into the egg, connecting it more fully with the atmosphere. Air is pulled into the lungs of the chick, compressed, wet, dogged, but new. This oxygen boost empowers the chick, gifting it the energy it needs to continue its struggle. It pecks at the weakened edges of the hole now, rotating its body and pushing its shoulders and legs firmly against the sides of the shell to provide extra anchorage. The beak continues to work across the roof of the egg, opening up this, its gateway, to the chick's new world.

Once the chick has climbed out of the shell, the adult bird, mother or father, deftly picks the eggshell up in its beak and drops it elsewhere, to remove its smell from the nest and deter egg predators. A short distance from the nest in which it was first placed, the eggshell sits on the floor, empty, alone and forgotten, the calcium atoms ready for another lap of the rock cycle. In time, perhaps, one in a million fragments of eggshells like these may become a fossil. Only a tinier fraction still will ever be seen or collected by human scientists. Most fossil egg fragments will erode to dust.

Much of what we know about the Cretaceous colonies of the so-called opposite birds come from fossils found in western Romania, specifically the Sebeş Formation of Transylvania. This spectacular layer of Cretaceous mudstone, announced to the world in 2012, was found to contain thousands of fragments of fossilised bird eggshell. In fact, 80 per cent of this accumulation consisted of smashed up, compressed layers of the stuff. One small subsection alone, less than the size of the open pages of this book, contained 150 fossil eggshell fragments. The skeletal remains of opposite birds from this and other Cretaceous sites show hatchlings with strong, calcified bones, wing feathers, big eyes and a large brain. These were chicks that were ready to fly within hours of their life beginning, suggesting a land-scape in which the threat of predation was high: the sooner birds could get up to move around, the better their odds of survival. We see a similar pattern in hatchlings of mudflat birds today, including ducks, geese and waders.

I find it incredible to think that every single fragment of egg from the Romanian fossil assemblage had probably been seen and inspected by a parent bird, lost to history. It is exciting to con-sider that each one was generated in an unknowable egg tube; that each egg contained calcium atoms stripped from that forgotten Cretaceous world; that each species was a unique trial, a new evo-lutionary expression of egg-ness. Within a few million years, much of that expression would be curtailed, cut short, snuffed out by a random rock from outer space that grazed our planet's surface.

The atmosphere changed immensely after the meteorite explo-sion that occurred 66 million years ago. For months, a choking sheet of dust painted the sky, preventing the sun's light from reaching the surface of oceans and continents. In the seas, phytoplankton were starved and their photosynthesis went into a terminal decline.

In their trillions, these life forms sank to the sea floor, dead. Some became cysts and acritarchs, but most did not survive. Food chains became incredibly simple for a while, almost like in the Ediacaran Period, hundreds of millions of years before. On land, plants withered and died, starving herbivorous animals, particularly the large dinosaurs. Those animals that depended on fish for their survival, some pterosaurs, some birds, found little to support themselves. There was food for those wily enough to locate it, of course: those able to home in on the dead bodies of others may have eked out a living for a while, for example. Perhaps months, perhaps years. But mostly there was death. For a short time, in geological terms, it would have looked like a dying planet.

The statistics for this mass extinction event are comparable to the end-Permian extinction crisis. Three quarters of animals and plants lost; almost every animal over 25 kilograms dead. This was the end of the road for the non-bird dinosaurs and for the opposite birds as well, although a fraction of 'true' birds survived and repopulated, in time. One day their eggs would find their way back into scrapes, into trees and the tallest cliff-side mountain tops. And these bird survivors would diverge into new lineages – the perching birds, the ground birds, ducks and their relatives and the ground-living fowl.

And what of the non-bird dinosaurs, those reptile rulers? Traditional hypotheses for their end include the fact that dinosaurs, many of whom were larger, probably had greater energetic needs and simply could not find enough food after the meteorite impact. Other suggestions for their disappearance are that, somehow, dinosaurs lacked sufficient adaptations for warm-bloodedness. That they perished in the cold climate brought about by the layer of dust that enveloped our atmosphere and shaded out the sun's rays. These ideas are freely expressed in dinosaur documentaries and popular science books on the subject. More recently, a new hypothesis has entered the mix, however – one about eggs. Microscopic analyses of fossil dinosaur embryos, specifically counting the growth lines on teeth as if they were rings in a tree, show that dinosaur embryos took far longer to develop in their egg than the embryos of birds – sometimes six months or more. This finding lends weight to a new hypothesis for the decline of dinosaurs: that the long incubation time of dino-saur eggs added 'lag' to their re-population rates in the years after the meteorite impact. Because of their slow-growing eggs, it may have been that dinosaurs competed poorly against their rivals, the birds and mammals, which nudged them out, edging them ever closer to extinction. It's possible that the dinosaurs may have remained to this day were it not for an apparently unsurmountable complication in how their eggs developed. We (the mammals) should be very grate-ful for this. We might never have had our time, otherwise.

Today, the birds that live in our neighbourhoods and that dot our skies continue to express the same forms of playful evolutionary creativity that have been their craft for more than 100 million years. The same journey out of the same egg tube; the patterns, dapplings, bruisings and pencil marks, noted and recognised; the same shell, cracked from within; the first connection between albumen-stained chick and weathered, tired-looking parents. Every song they sing each spring is an ode to more.

Long may that continue.

Interlude
A post-Cretaceous moment

I n the aftermath of the catastrophic collision, a once-thriving world was transformed. In a matter of moments, the world's lush forests were replaced by a desolate, ashen kind of terrain. Skies became perpetually overcast. A gloomy, ashen fog bathed all things for years on end.

Here, somewhere on the ground, among rocks and fallen branches, lay *Tyrannosaurus rex*, the last of its kind ever to exist. Struggling for breath, starved and listless, it succumbed to a lonely death. It was over surprisingly quickly.

On the first night, the body of that enormous dinosaur rested alone, unnoticed and untouched by anything. This isolation would not last long, however.

The next morning, the sun rose. The morning brought with it a gentle wind that carried the smell of dinosaur decomposition high into the air, where it was picked up by an insect. A fly, something rather like (but not quite) a bluebottle. This insect followed the scent to the fallen archosaurian predator.

The fly landed on the dinosaur and, within a few seconds, began to explore its folds. For minutes on end, the fly meticulously surveyed every contour and every crevice of the colossal dinosaur's form, each cryptic orifice, every subtle detail of its leathery skin. The fly's movements would have looked mechanical, almost robotic, had you been able to watch. With purpose, it walked across the body of the collapsed giant, searching for a suitable location to lay its eggs.

Within a few minutes, the entire dinosaur explored, the fly returned to the first spot it had landed upon. The corner of the dinosaur eye socket would turn out to be a fitting location for eggs, a damp place safe from wind and sun. And so, in the minutes that followed, the female fly commenced the measured task of laying its eggs. One, two . . . five . . . ten . . ., each egg was tucked in close to the inner lining of the eye socket, each egg packed tightly against one another to limit exposure to the frigid atmosphere. The eggs looked shiny, waxy, clean, ordered.

The female fly did not deliver all of its eggs in one go. Every few minutes or so, it buzzed to a nearby perch, where it meticulously groomed its legs and egg-laying anatomy before returning to deliver another round of eggs onto the *Tyrannosaur*'s closed eyelid. And then, an hour later, after a few more rounds, the fly was finished. Truly finished. For the effort of egg laying proved too much for the fly and so the short-suffering insect promptly ceased to be.

In the long shadow of the fly's meticulous work, the *Tyrannosaurus rex*, once a lifeless canvas, became a nursery of hope and renewal. And in that moment, as the stitches in time crossed, through the medium of eggs, the tapestry of life was beginning to renew itself.

11

THE INVASIVE PLACENTA

Cenozoic Era, 66 million years ago to today

'What if we did come where we are by chance, or by mere fact, with no one general design?'

— William James (1842–1910)

Seventy thousand years ago, in the edge lands of a continent that then had no name, in a forest clearing on the side of a mountain, in the light of a fire, knelt a grandmother who had seen it all before. Through direct experience, through memories of others, through conversation, through stories, the grandmother knew birth. This elder knew how to calm an expectant female. They knew the position that best eased the neonate from its birth canal. This was a collective wisdom, built on predecessors, that echoed down the ages. Mother after mother, grandmother after grandmother, bringing new life to the world.

The grandmother's steady hands, calloused by years of finding and preparing food, moved with practised grace. This familial guide knew the rhythm of life intimately, the routines of contractions, the groans and the wails of a pregnant ape. But the grandmother was also as much a product of evolution as the female giving birth in front of her. This was an animal for whom menopause, a rarity among mammals, had evolved. A lengthier life had resulted from this adaptation – one in which more experience could be accrued, more experiences shared, more lives brought safely into the world. Fledglings of these family groups formed care-giving communities, built upon the sharing of knowledge, the passing of genetic material their unknowing goal.

The birth is, in the end, one of the quicker ones. Immediately upon entering the world, the neonate's shrill screams fill the air. The mother, living, breathing, both awake and not awake at the same time, clutches the young mammal to a nipple. The baby feeds. The umbilical cord, chewed clean through, is then wrapped around the defenceless newborn. And then something else moves into the world. The placenta is expressed silently out of the uterus and, with that, the life-giving interface between mother and offspring in the womb passes. In this warm, life-filled place, perhaps the most energy-sapping placenta the world has ever seen silently dies.

In this, our final chapter, we'll come to understand how this extraordinary adaptation became somewhat super-charged in the human lineage. But first, some context is needed, namely the placing of our placenta in time – the Cenozoic Era, sometimes called the Age of Mammals.

Part of what we know about the early advance of mammals after the demise of dinosaurs comes from a single fossil site, the Messel Pit, found near Frankfurt, Germany. This pit, 1 kilometre across, is partially filled with water now. Its banks are sectioned and square, tiered like the seating plan of a heavenly auditorium. It is a land of rubble and rockfalls, colonised by secondary woodland. But still, here and there, the black, oily shale protrudes from out of the rock to reveal the lives of animal dramas past. Here, in the fossil layers of the Messel Pit, the level of preservation is nothing short of spectacular.

Messel is located in what was once, 50 million years ago, an Eocene lake surrounded on all sides by subtropical forests that were spectacularly rich in animal life. Snakes, lizards, amphibians, insects, crustaceans – all are preserved with their finest details still intact. This stunning preservation is down to the fineness of the sediment that once ran through this landscape. Much of it consisted of ancient mudfalls, brown dusty slurries, whose soft sediments gently cloaked dying animals in the water, preserving layers of bodies like pages in a photo album. Bacteria had barely enough time to act upon many of these organisms, so quick were the tumultuous floods which occasionally entombed them. And so, because of all this, the Messel Pit has become one of the finest windows through which we can view how quickly, within 15 million years of the death of the largest dinosaurs, life bounced back.

Messel slabs show giant ants, flying termites and jewel beetles, their iridescent blues and greens still clearly visible. There are spiders pulled from webs in moments of tragedy; bowfin fish and garpikes and eels stare ghoulishly outwards, as well as turtles, two preserved in coitus. There are crocodiles of six species and constricting snakes and legless lizards. Birds of many kinds are present, some evolving into forms we might recognise today as owls, flamingos, swifts, nightjars and rollers. There are long-tailed, spiny shrew-like mammals that propped themselves up on their back legs; rodents, with charismatic long incisors at the front of upper and lower jaws; there are bats, early pangolins and hyena-like predators, some barely the size of weasels; there are also semi-aquatic otter-like mammals that fed upon insects. There are anteaters that resembled the giant anteater that still lives in South America today and distant relatives of animals that would one day evolve into rhinos and tapirs. And then, among them, lemur-like animals that were early trials in primate evolution – the distant ancestors, perhaps, of us.

The ancestors of horses, particularly, lived in abundance back then. Eight *Eurohippus* fossil specimens show within them the remains of young at varying stages of development. Some look as if they are just days away from birth. At Eckfeld, in a closely related species named *Propalaeotherium*, the sediment has preserved many more of the soft tissues. It is in one adult early horse specimen that you see it: a foetus crumpled within a cage of fossilised ribs, its tiny legs curled around its short, downward facing head; and then, next to it, a faint merging with something dark and inky. It is a fossil of a mammalian placenta, the finest preservation of this organ known. X-rays and scanning electron microscopy imagery provide an enhanced view of the organ's soft tissues upon which can be seen a lawn of fossilised micro-organisms that were, in life, beginning to consume the rich spoils as the decomposing horse sank further and further into the mud.

It is well known that mammals came to dominate the niches left after the demise of dinosaurs; but it is the diversity of their forms that is most striking of all. Each continent, long split from Pangaea, spawned its own mammalian shapes and arrangements. South American forms, detached from the rest of the world, were particularly spectacular. Toxodon, a herbivore the size and shape of a rhinoceros, with a head like a spatula, was called by Darwin 'perhaps one of the strangest animals ever discovered'. There were South American mammals that filled the niches held by hippos, camels, rabbits and deer across Africa and Eurasia today. Ancestors of sloths that were the size of elephants stood upon their hind legs, scything at low-hanging leaves and fruit – fragments of their pelts are still found in caves today; in some there are still scratch marks on the walls. Rodents, a group now numbering almost 5,000 species, attained the largest of sizes during this period, becoming almost bear-like in some cases. The name of some of these rodents, the 'dinomyids' (meaning 'terrible mice') says it all. Armoured giants like glyptodont sloped across the pampas; their close relatives live today as armadillos. In the south, these animals thrived. In the northern continents lived the ancestors of bears and wolves, elephants, hedgehogs and others known from the Messel Pit fauna. There were whales too. In fact, it took just 25 million years for them to assume the familiar shape we see in them today. The earliest whales were the size of wolves; they had functional limbs and could live amphibiously, hunting in water while mating, playing and resting upon the land. One, a species known as *Maiacetus*, is described from a fossil in which mother and baby, an unborn foetus, are clearly preserved – the species' name translates as 'mother whale'.

It is incredible that, for each evolving mammal lineage, there would have been the push and pull of evolutionary deals between mother and offspring, leading to ever-greater diversity in placenta shape and size. Some placentae were disc-like, some leaf-shaped, ring-shaped, heart-shaped; some placentae were internally folded; some ridged, branched, labyrinthine, web-like.

In scientific literature, the elephant placenta is particularly well studied. Up to 20 kilograms in weight, it has upon it rings of villous tissues, with ridges and girdles. The umbilical cord, as thick as a hose, can be more than a metre in length. Unusually for land mammals, the elephant placenta is rich in myoglobin, a muscle protein that binds reversibly with oxygen. The only mammals with such high myoglobin levels are diving mammals, including dugongs and manatees (collectively known as 'sirenians'). These mammals use placental myoglobin as an extra source of oxygen when the pregnant female is submerged in water, holding its breath. The high levels of myoglobin in elephants suggests that, deep in their history, their ancestors almost certainly lived a semi-aquatic lifestyle.

Less well studied, but more commonly seen, are the placentae of whales. Sometimes they are spotted floating in surface waters, often by people on whale-watching boats that happen to get too close to whale birthing grounds. From a distance, the floating sofa-sized placenta resembles a mat of toilet tissue floating on the surface of the water. Up close, the umbilical cord, as thick as a gutter pipe, snakes lifelessly.

Although it has not been preserved in the fossil record, the diversity of placentae among modern-day mammals suggests that, about ten or twenty million years after the end-Cretaceous, at around the time that the animals of the Messel Pit were alive, the mammal placenta was changing. Natural selection was tweaking this organ. In many cases, it was selecting the individual placentae best able to extract as much energy from the maternal host as possible. Yet, surprisingly, in some lineages the placenta appeared to take a step back, becoming less, rather than more, invasive. Looking at data across sixty mammal species, a trend becomes apparent. Plotting the invasiveness of each placenta (judged partly by how many blood-gathering finger-like projections the placenta has) against important life-history details, such as how long a species takes to mature and how many offspring each year a species might produce, the least invasive mammal placentae are those associated with a more rapid pace of life. Species that live fast and die young, in other words, appear to end up evolving a less invasive placenta. One argument for why this might have been goes as follows: it begins with a change in the environment, say the loss of a food resource. In this situation, as the probability of death increases among the population, natural selection favours those mothers that produce offspring at a younger age – the quicker you can reproduce, and the more offspring you have, the better the chance that at least one will survive. In this resource-limited world, made worse because mothers have less time to accrue resources before producing offspring, the mother becomes, to put it in financial terms, strapped for cash. The mother has fewer resources to invest in reproduction and so the placenta adapts to this change in circumstance, lest it 'bankrupt' its mother. In other words, when an embryo is made cheap, it loses some of the bargaining power we saw at play during the Jurassic. This change in life-history strategy may, essentially, liberate the mother from being manipulated by its offspring.

Brain size is another marker that tracks closely with how invasive a placenta evolves to be. Not just how *large* the brain is in relation to the body, but also how *quickly* the brain grows before birth. Both factors correlate with especially invasive placentae. How it works is simple: the bigger a mammal's brain evolves to be, the greater selective force is placed on the placenta to acquire energy for the growth of the embryo, which, naturally, drives the evolution of an ever-hungrier placenta. In some big-brain lineages, our own included, this causes serious complications (which we will soon discuss) that burden many pregnancies to this day.

Mammals are, as a group, more brainy than other similar sized organisms, but this wasn't always as key a feature of our kind. It seemed to happen gradually, after the demise of dinosaurs and as the Cenozoic Era began to progress. Scientists had originally thought that this relative increase in brain size was simply a by-product of the evolution of a larger body size in mammals, but recently (using three-dimensional models of fossilised mammal skulls) this assumption has been more rigorously tested. At first, it seems, in the 10 million years after the era-ending meteorite, mammal body size increased and, relatively, so too did brain size. But then, clearly visible at fossil sites like Messel, brain size in certain lineages increases at a higher-than-expected rate compared to body size. Mammal brains, in some lineages, were given a metaphorical shot in the arm. So why? If they cost both mother and foetus more to produce, particularly in the embryo stage, what's so good about big brains?

The researchers who first made this observation about brain sizes in mammals, comparing three-dimensional models of fossilised skulls, think that this trend occurred because of competition. At first – without dinosaurs and other large land animals – plants, insects and other resources were easy to harvest and competition between individuals was low. In this environment, energy-sapping brains were costly and unnecessary. But later, when mammals diversified and established themselves – when there was more competition for niches, for food and resources – smarter individuals were comparatively more successful in some species. In terms of the transmission of genes, large brains began to pay out and, in some lineages, bigger and better brains started to evolve. In some mammalian groups today, such as dolphins, rodents and particularly primates (monkeys and apes), the ratio of brain size to body size has continued to increase with time. In humans, perhaps the wiliest of all primates, the trend continued with aplomb.

There is no denying the selection pressure at work here: big brains really are inordinately expensive for bodies to build. And human brains truly differ from the brain of our nearest relatives, the chimpanzees. At birth, for instance, a chimpanzee's brain is 130 cubic centimetres and then triples in size in the following three years. Compare that to the human brain. At birth, the human brain is more than *twice* the size of a chimpanzee's and, in six years, it *quadruples* in size. Although our brain takes up just 2 per cent of our total body weight, this organ consumes between 20 per cent and 25 per cent of our resting energy budget. The human brain costs something like 420 calories a day to run, four times more than the chimpanzee's brain.

This is why the relationship between human mother and child, connected via the placenta, has become, evolutionarily, so strained in our species. More strained, it seems, than in any other mammal.

Liam Drew, author of the authoritative *I, Mammal*, points out exactly how twisted this relationship becomes. For starters there's pre-eclampsia, when the mother's body goes through a life-threatening surge in blood pressure as the human foetus increases the rate of blood flow through the placenta. Put simply, it wants to be bathed in as much life-giving blood as possible. And there's gestational diabetes, caused by the foetus' attempt to co-opt maternal control of blood sugars – predictably, it wants more than the mother is able to give.

Pre-eclampsia affects about 5 per cent of human females carrying a single baby to term. Add more offspring into the mix, twins or triplets say, each of whom will often have their own placenta, and pre-eclampsia rates increase to one in three pregnancies. This makes childbirth a risky activity for human females.

There are other tricks that the placenta has evolved to get what it needs for the embryo. Staggeringly, we now know that the placenta uses a special protein (called PP13) to inflame the tissue around tiny veins in the uterus, causing the mother's immune system to invest heavily in immune defences. It's a classic distraction technique evolved by the placenta: if the mother's immune system is firefighting elsewhere, it is less likely to focus its attention on combatting the placenta's active uterine encroachments.

What results from all of this, says Cat Bohannon, author of *Eve: The Real Origin of Our Species*, is a 'nine-month stalemate': 'women's bodies are particularly adapted to the rigors of pregnancy not simply so we can get pregnant but so we can survive it,' she writes.

The highly invasive human placenta, influenced by our enormous brain and (probably to a lesser extent) our slow-and-steady life history, also explains another quirk of our species, the phenomenon of menstruation. This adaptation is fleetingly rare among mammals, found only in some primates, bats and elephant shrews. In humans, menstrual bleeding is particularly overt and, by now, after reading the previous paragraphs, you can probably guess why.

Having an extra-thick uterine lining helps the female survive the potentially hostile tentacle-like villi of the placenta, should pregnancy occur. The lining of the uterus in our species has become so thick that we cannot possibly reabsorb it every few days or weeks, as other mammals do. It is more efficient, in our species at least and a handful of others, to shed the uterine armament and grow it afresh each cycle ready for the next potential implantation.

And so, human evolution has occurred both due to, and in spite of, the placenta. Every pregnancy, unthinkingly, must navigate a careful path through it. Every menstruation is testament to it. It is partly why menopause exists, to give individuals an escape from the energetic costs associated with its imposition. This life-history phenomenon only exists in a small number of apes and some whales and dolphins.

In many years of writing about the insides and outsides of animals, I confess I have never written of a stranger organ or a weirder evolutionary contract. I find myself quietly saluting the placenta that fought for me in my earliest moments, while simultaneously feeling apologetic to the maternal host in which I grew. This is a world-changing adaptation, in more ways than one.

And so, in the Cenozoic Era, eggs became dotted across continents in a very tribal way. The eggs of amphibians, mostly, in water; the eggs of reptiles – mostly soft-shelled – in the soil, in the sand, in leaf-piles, among the nooks of tree roots. The birds, with their calcified and decorated eggs, deposited them in forests, grasslands, shores and cliff tops; the mammals, wedded to embryos shielded from predators and protected from the extremes of our atmosphere, became one more adaptation in a continuum of how eggs are made, nurtured, cared for by the animal body.

It is a shame that life could not stay this way; that the forms and diversity that arose from out of the ashes of a meteorite could not continue to blossom a little longer before the wheels of extinction gathered pace once again. That a bushy lineage of upright primates, evolved in Africa, would come to shake dramatically the distributions of eggs on our planet would have been very hard to predict from the vantage point of the Cretaceous. Yet it took less than 70,000 years for our species to change the face of almost every single continent that exists today. And so, the Cenozoic will end as an age when the egg was employed as an agent of change. By understanding animal reproduction, this became the age of cattle; the age of chickens; the age of you and me. We learned the secrets of animal eggs and, in a blink of geological time, less than a freeze frame on a flaky slab of fossil-rich mudstone, so many things changed.

Today, there are more than 8 billion of us. In mere centuries, three quarters of our continents have been significantly altered by our actions, principally through farming; two thirds of the oceans are affected by fisheries and human-caused pollution. Shaped by our reproduction, particularly by our large brains, we now find ourselves (whether we like it or not) in the driving seat. We are the first organisms in the history of the planet to modify the constituents of land, in part by choosing which animals reproduce and which do not. As you read these words, 60 per cent of all mammalian biomass comprises farm animals – we account for 34 per cent. Just 4 per cent of mammalian biomass consists of wild animals – whales, lions, rats, anteaters, sloths and more, so much more. Statistics like these are nothing short of extraordinary.

Chapter by chapter in our story of eggs, we have seen how adaptable the egg is in the face of environmental change. How the presence of storms, solar radiation, the drying atmosphere, forest fires and meteorites from space have seen the egg strengthen, adapt and enhance itself many times over. It is no surprise that we are seeing this happen again, in this, the modern part of our evolutionary tale.

Today, because of our changing climate, the eggs of insects, predictably, are adapting most quickly. In the UK, for instance, butterflies and moths hatch up to six days earlier than they did just ten years ago. In the USA, insect activity begins twenty days earlier than it did seventy years ago. Aphids, considered a pest upon trees and other plants, now hatch from their eggs a month earlier than they did half a century ago. Fast generation times and variation between insect populations is seeing natural selection work quickly in these species. Insect distributions are also altering because of human-induced climate change. Traditionally, cold winters (which kill off insect eggs) were a natural barrier to the movement of invasive species, but milder winters have seen their spread across countries and continents continue apace. This is why, entomologists argue, mosquitoes have brought new diseases to the European continent in the last decade, including dengue fever to France and Croatia, chikungunya in Italy and malaria in Greece.

Today, mild winters are affecting other problematic invertebrates too. In the USA, cases of Lyme disease, spread through blood-sucking ticks whose eggs are no longer killed off in the winter months to the same degree, have increased three-fold in twenty-five years. There are similar concerns over populations of tree-munching ash borers and the invasive Asian giant hornet, a potential threat to honeybee populations across Europe and North America. As they have always done, invertebrate eggs are adapting quickly, unthinkingly profiting from the climate crisis.

Although their life history plays out at a slower pace, the life-history patterns of land vertebrates have also begun to shift in recent decades in response to shifts in climate. Information on birds, which have a long history of amateur study, best show the changes. Most notably it is very clear that, like insects, many bird species are adjusting the moment at which their eggs are laid and at which point they hatch, potentially to align themselves better with the seasonal availability of food. In North America, roughly a third of birds lay their eggs earlier, by about twenty-five days, than they did a century ago. In the UK, between 1971 and 1995, 63 per cent of bird species nested earlier, by an average of 9 days.

Perhaps a more pernicious danger to vertebrate eggs today is that posed by extreme weather events. In sea turtles, for instance, we may point to warmer temperatures and thus more of their population turning female (a strange quirk of turtle embryos is that their sex is determined by temperature), but unpredictable storms and the flooding of shoreline nests is perhaps the far greater risk. Hurricane Irma in 2017, for instance, saw 56 per cent of green turtle nests and 24 per cent of loggerhead turtle nests on the coast of Florida destroyed. In 2019, Hurricane Floyd killed 100,000 turtle hatchlings in one fell swoop. The truth is that, for many species, there may not be time for the egg to adapt to environmental changes so sharp and jagged as this.

There are likely to be both many losers and perhaps a few winners in these turbulent times. The wolf spider, *Pardosa glacialis*, may become one of those that benefit from climate change, for example. In 1996, this tundra-living spider used to lay one clutch of eggs each year, but now it takes advantage of a longer spring and summer and regularly lays a second clutch of eggs later in the season, often with more eggs than the first. Because the spider has more months in which it can hunt, the size of adult spiders at the end of their season has also increased. The significance of just this one change to an ecosystem could be far reaching: a single square kilometre of tundra is home to one million of these ground predators – spider-eating birds, perhaps, have a lot to gain. Finding a way to study these populations, to record ecosystem shifts happening due to climate change, will prove crucial to the human response.

For decades, museums have been one of the best tools we have to measure the impact that the climate crisis is having on eggs and the organisms that hatch from out of them. Globally, there are specimens, records and data from five million bird eggs collected from as far back as 250 years ago. Eggs are different to other museum specimens. Carefully stored, they do not rot; they require very little by way of preservative fluids and, within each specimen is a chemical record, through fragments of DNA, of breeding biology and diet. Nowadays eggs can be scanned with electron microscopes and spectrophotometers; they can be genetically sequenced or scraped of their isotopes for carbon dating; they can be scanned and stored and shared in a digital format. There may be no more perfect kind of specimen in the world from which to gather data about our changing planet.

It was museum eggs, of course, that first allowed scientists to prove the link between heavy DDT (dichlorodiphenyltrichloroethane) use and declining bird populations, a story detailed so eloquently in Rachel Carson's landmark *Silent Spring*. By comparing eggshells from the late 1940s to the 1960s, when DDT use was high, with museum-curated eggshells before this time, scientists were able to show that DDT was pooling in birds, particularly those higher in the food chain such as birds of prey. This caused them to lay easily broken eggs with thin shells, causing wild populations to decline. The result was the agricultural use of DDT being banned in most developed countries. Over the years, other studies have seen environmental changes linked to eggshell thinning. Most notable are those that firmed up the causal link between 'acid rain' and eggshell malformation in birds. Studies have shown a long and drawn-out reduction in eggshell thickness in blackbirds, song thrushes and mistle thrushes, for instance, all linked to acidification of their lowland environments, mostly through pollutants such as sulphur dioxide and nitrogen oxide. The main problem appears to be that acidification, which removes calcium carbonate from the soil, leads to a reduction in the number of snails, whose shells provide calcium for nesting birds readying themselves to lay. In each case, museum specimens act as the vital comparison group for the data that scientists collect from the field. Eggs may, in part, help us better understand environmental changes, but the clock is clearly ticking. Kyoto, Copenhagen, Paris, Glasgow: the agreements arising from global climate conferences, useful as they are, have so far failed to budge carbon dioxide emissions from their upward trajectory. Eggs cannot arrange summits or conferences and they make lousy politicians, but, I suspect, if we observe them more closely, take them more seriously, celebrate them where we can, they would have plenty more to tell us about how and why we might limit the impacts of the raging climate change we are now inflicting upon our world. That an extinction crisis is approaching is obvious. The question is becoming, at this rate and trajectory of change, what kind of world do we want on the other side, in the next chapter? What level of impoverishment will we tolerate? How much suffering are we comfortable with?

The mud and silts laid down today will record the trials and evolutionary experiments of new eggs, as they always have. Baked and dried, hidden for millennia, the fossil strata will become the pages upon which the story of eggs is written in future chapters. As long as our nearest star shines, I have no doubt that there will be eggs on Earth. Their journey will never be finished. I hope we will be around as long as possible to see it continue.

The invasive placenta, the adaptation that in many ways defines our species, has seemingly hit upon a way to carefully negotiate an uneasy, yet stable, relationship with its maternal environment, one where both sides live another day. It may be that we humans end up drafting a similar contract with the planet. A relationship where our big brains are sated through the environment, where they can be nurtured and prosper, but not at the expense of the maternal vessel that carries us forwards in time, onwards through space.

Evolution frequently sets down challenges which life, and eggs, evolve to counter. Whether we are capable of the same is still to be seen.

EPILOGUE

A future for eggs

Three hundred kilometres out to sea bobs a white plastic bottle. It rolls upon each gentle wave and dips over each crest. The lid, firmly attached, points windward like a figurehead on the front of a miniature warship. The frayed label, almost completely perished, licks at the breeze like a sail. There is no grandness to the object, so thoughtlessly discarded, yet this bottle is actually covered in life. The white texture upon its surface is due to a film of 70,000 tiny eggs, each half a millimetre or so in diameter. The eggs have not finished coming, either: more are being laid by a cloud of opportunistic animals – they are ocean-faring, sea-skating insects. They hustle and climb and writhe against one another in a disorderly, highly localised, plague around the floating bottle.

Every minute or two, more seem to arrive. There is no space on the bottle for these late arrivals and so the insects laying their eggs simply attach them in new layers on top of old. They scramble and pile over one another in a state of desperation. Their eggs must be laid here, or else they will die.

In the surrounding waters, within a metre or two, there are riotous aggregations of unpaired male insects, seeking more females with whom to mate. Like figure skaters, the male insects dash and weave on the surface of the ocean, actively trying to sink their rivals or chase them from this, their temporary territory. Not a single sea skater appears deterred by the baking sun high above; each is uninterested, apparently, in the hungry seabirds drawn from far away to this cloud of activity. All that matters to each insect here is the plastic bottle, their floating Rock of Gibraltar, this island of eggs.

In the days that follow, the dazzling Anthropocene sun continues to blaze down upon these eggs; by night, they glint and glisten towards the star-kissed face of our planet's expansive moon. Beneath it, the eggs do what eggs do: they generate the potential of more animals. Inside, cells divide. Two becomes four; four becomes eight; eight becomes sixteen, and so on. New lives form in these tiny spherical worlds, readying themselves for the next.

Within each egg, an embryo will develop that will eventually have its moment in the narrative of existence. And when this organism finally hatches from its egg, here in the middle of an ocean on a planet third nearest to its star, it will have the simplest of evolutionary aims. To make more. That make more. That make more.

ACKNOWLEDGEMENTS

First, I would like to extend enormous gratitude to everyone at Elliott & Thompson, particularly editors Sarah Rigby and Pippa Crane, who have been nothing short of instrumental to the building of this book. Through their patient edits, both small and not so small, I got more out of the writing process than I ever thought possible. Also, I am very grateful to Nik Prowse for many helpful suggestions.

My thanks go out to the following readers, who helped me steer the text and guided me away from the superfluous: Hana Ayoob, Ben Barlow, Jane Barlow, Josh Luke Davis, Lawrence Foster, Neil Goodwin, Adam Hart, Dave Hone, Nicola Khan, Phil Lashton, Nick Linstow, Chantal Lyons, Phil Martino, Emily Millhouse, Jessica Pollitt, Jon Samrai and to everyone at Froglife (www.froglife.org), of whom I am a dedicated supporter and patron.

And for all their support and kind words on this book and others, giving me confidence with writing when sometimes I despaired: Lucy Cooke, Liam Drew, Jess French, Ben Garrod, Melissa Harrison, Ben Hoare, Kate MacDougall, Erica McAlister, Helen Pilcher, Alice Roberts, Helen Scales, Richard Smyth, Isabel Thomas, Zazie Todd, Laurie Winkless and Becky Wragg Sykes. I am enormously grateful to Ruth Kent, my loving and loyal cuttings provisioner for more than a decade, and an early reader of this book.

Special thanks to my fantastic agent Laura Macdougall of United Agents, both for inspiring me to write a bit differently and for believing in eggs from the very first discussion that we had. Thanks also to Olivia Davies and Emily Talbot, also of United Agents, who have also given so much help and support over the years.

I am also very grateful to Tim Birkhead for his book *The Most Perfect Thing: Inside (and Outside) a Bird's Egg*, an instructive, authoritative, yet personal work. Quite simply, this book would not exist without it. If you have not read it, I recommend that you do.

This book couldn't have been made without my parents, who have always offered so much love and support. This one is dedicated especially to my mum. Every cell I have contains the mitochondria you provided. Thank you for everything. You (and Dad!) really are an inspiration and I've learned a great deal from you about life and love.

Finally, to Emma and Scarlett and Esme – my Big Three. Thanks for giving up part of the kitchen table for me and sorry for all the books lying around for what feels like an eternity. Thanks for all the supportive words and the handmade cards stuffed with motivational sentiments. Most of all, thank you for inspiring me with your love of stories and your love of words. Books are eggs too. They contain almost all of what is needed for life. I wouldn't want to share them with anyone else.

GLOSSARY

ACRITARCH — a broad term used to describe unicellular fossilised marine planktonic organisms, including resting stages (cysts).

ALLANTOIS — a membrane of amniote eggs (reptiles, birds and mammals), formed as a pouch from the hindgut. In placental mammals, the allantois closely associates with the chorion in forming the placenta.

AMNION — the membrane surrounding a developing embryo.

AMNIOTE — an animal whose embryo includes an amnion and chorion and has an allantois — specifically, a mammal, bird or reptile.

ARTHROPOD — a joint-legged animal of the phylum Arthropoda, which includes insects, spiders, crabs, mites and scorpions.

BLASTULA — the early stage of development when the embryo becomes a hollow ball of cells.

BROOD PARASITE — an animal that lays its eggs in another nest. A term used mostly in the context of birds, but isolated examples occur in fish and mites.

CALCITE – the mineral form of calcium carbonate, common to mollusc shells.

CHORION – the outermost membrane that forms around the eggs of many animal groups, most notably in amniotes. In mammals, it is the chorion that contributes to the formation of the placenta.

CLOACA – The posterior opening in reptiles, birds and amphibians, from which urine, faeces and eggs are ejected.

CNIDARIA – the phylum-level group name to describe jellyfish and their allies, including *Hydra*.

CORTICAL GRANULES – small packages found in oocytes that regulate and coordinate actions across the cell membrane during fertilisation.

CYCAD – a group of palm-like shrubs, common during the Mesozoic Era.

CYST – the resting stages of unicellular planktonic organisms. Also used to describe hard-wearing eggs of fairy shrimps.

EURYPTERID – the group name for sea scorpions, early relatives (likely a sister group) of spiders and other arachnids.

GASTRULATION – the stage during early embryonic development when the blastula (a hollow ball of cells) is reorganised into a multilayered three-dimensional structure known as the gastrula, from which development continues.

GENUS – a classification group, one level above species level. The horse (*Equus ferus*), for instance, is in the genus *Equus* – the group containing donkeys, horses and zebras.

GERM – the name given to cell lines that go onto produce sex cells, rather than body cells, and are thus able to directly pass on genes to future generations.

HOLOMETABOLOUS – a descriptive for insects that undergo 'complete metamorphosis' – eggs, larva, pupa, adult.

INVERTEBRATE – a broad term used to describe any animal that lacks a backbone.

MARSUPIAL – pouched mammals, commonly found in, but not exclusive to, Australasia.

MITOCHONDRIA – sausage-shaped organelles found in plant and animal cells where biochemical processes of respiration and energy production occur.

MOLLUSC – any member of the phylum Mollusca, which includes squid, octopus, bivalves and snails.

MONOTREME – the group of mammals that lays shelled eggs. Modern representatives are the platypus and echidna.

OVARY – the site of egg production in female animals. Paired in most species, except birds.

OVIDUCT – a general descriptive for the tube through which eggs pass from the ovary of a female animal.

OVIPOSITOR – the needle-like structure used, particularly by parasitoid wasps, to deposit eggs in hard-to-reach places.

OVULATION – the movement of the unfertilised egg from the ovary into the oviduct.

OVUM – the female sex cell. *Plural*: ova.

PALYNOLOGIST – the name given to scientists who study plant pollen, spores and microscopic planktonic organisms (collectively termed palynomorphs) that resemble eggs.

PANGAEA – the supercontinent that existed through the Permian and Triassic Periods.

PARASITE – an organism that lives within or upon another organism (the host) and benefits (through feeding or otherwise) at the expense of the host.

PERIOD (geology) – a sub-division of geological time (eon, era, period, epoch . . .). For instance, the Mesozoic Era is split into three periods: the Triassic, the Jurassic and the Cretaceous.

PLACENTA – an organ through which an embryo is nourished and maintained. In mammals, the placenta is connected to the mother by an umbilical cord.

POLYP – a cnidarian feeding structure, consisting of a simple mouth surrounded by tentacles.

PROTEIN – organic chemicals made up of long chains of amino acids, which form the building elements of life.

SEDIMENTARY ROCK – a rock that came from the laying-down of sediment, such as sand (sandstone) or mud (mudstone).

SEROSA – an egg membrane, common to most insects, that keeps the embryo from drying out.

SOMA – the name given to cell lines that become body tissue, rather than sex cells, and are thus unable to directly pass on genes to future generations.

SPERMATOPHORE – a vase-like structure, placed in the environment by males, which contains sperm.

SPONGE – primitive marine organism with limited tissue types.

STROMATOLITE – bulbous structure constructed, layer by layer, by cyanobacteria growing in slimy mats.

TETRAPOD – a term which describes the four-footed descendants of the earliest landfish. Amphibians, reptiles and mammals are all tetrapods.

TRILOBITE – an extinct arthropod group, once abundant, with a hard exoskeleton made from calcite.

UTERUS – the region of the oviduct in which the embryo matures, often called, loosely, the 'womb'.

VERTEBRATE – an animal with a backbone – fish, amphibians, reptiles, birds and mammals.

VIVIPARITY – the name for the reproductive strategy in which the embryo develops inside the female body, nourished by (in many cases) a placenta.

YOLK SAC – the membrane within which energy-rich yolk is secured.

ZONA PELLUCIDA – the membrane of the ovum that, among other things, regulates interactions between sperm and egg.

CHAPTER NOTES AND FURTHER READING

1. Dust from dust

Agić, Heda, 'Fossil focus: acritarchs', *Palaeontology Online*, 6, 1–13 (2016)

Al-Khalili, Jim, McFadden, Johnjoe, *Life on the Edge: the Coming of Age of Quantum Biology*, Crown Publishers (2014)

Demoulin, CF, Lara, YJ, Cornet, L, François, C, Baurain, D, Wilmotte, A, Javaux, EJ, 'Cyanobacteria evolution: insight from the fossil record', *Free Radical Biology and Medicine*, Aug, 206–223 (2019)

Fortey, Richard A, *Life, an Unauthorised Biography: a Natural History of the First Four Thousand Million Years of Life on Earth*, HarperCollins (1997)

Heusser, CJ, Palynology. In: *Paleontology. Encyclopedia of Earth Science*, Springer (1979)

Lane, Nick, *Life Ascending: The Ten Great Inventions of Evolution*, Profile Books (2009)

Rutherford, Adam, *Creation: How Science Is Reinventing Life Itself*, Penguin (2013)

2. The garden of mortality

For an excellent account of Namibia's landscape and fossil history around the Aar Plateau, I highly recommend Mark McMenamin's *The Garden of Ediacara: Discovering the First Complex Life.*

This chapter includes information on how jellyfish (specifically hydrozoans) have been in helping us understand the idea of the soma and the germ. While researching it, I spent a lot of time (too long, really) reading old embryology journals. There is, I'm sure, a book yet to be written that details the lives of the technicians and researchers who worked in the field of embryology in the nineteenth century, many of them women. The research facility in Naples, Statione Zoologica, was especially ground-breaking, setting aside a section of its research laboratory – a 'research table' – to be hired by any American research institution committed to seeing women involved in what was then a fledgling science. Within a few years of the Women's Table setting up in 1897, women became a mainstay of the research teams working at the facility. Among the first to arrive was Florence Peebles, who studied the regeneration abilities of marine invertebrates. Peebles' interest was in stem cells and how they contribute to Weismann's ideas of an apparent division between soma and germ cells. Then came Emily Ray Gregory, whose interest was in turtle embryos and the development of their reproductive organs in the earliest weeks of life. After that was Nettie Stevens, who discovered that some chromosomes determine the sex of a developing embryo: that some sperm carried large chromosomes (later known as the X chromosomes) and some sperm carried small chromosomes (Y). This discovery provided important support to Gregor Mendel's then recently unearthed manuscript, which outlined the passing of 'genes' from generation to generation as the mechanism through which inheritance occurs. Stevens' is one of the greatest discoveries in the history of science, although many are unaware of her name.

Another name that rarely finds its way into the history books, who regularly took a seat at the Women's Table in Naples, was the embryologist Ethel Browne Harvey, who worked on the actions of stem cells in *Hydra* and undertook experiments on the chromosomes of sea urchins, again underlining the mechanisms of inheritance. Harvey's influence on embryology is so strong, it is argued, that she was instrumental in the awarding of Nobel Prizes to not one, but three male scientists. But there are others too. Mary Dorothea Gruber, the wife of August Weismann, who observed, detailed and relayed information to her husband, then sight-impaired, about the organisms prepared on microscope slides. Marcella Boveri, the 'Zoological Miss' — wife of the so-called founder of cell science Theodor Boveri; Miriam Menkin who, in 1944, while practising techniques for *in vitro* fertilisation, became the first person to conceive human life outside of the body; Jean Purdy, the embryologist involved in the conception of the first IVF baby, Louise Brown. These people really matter in the history of this science.

Adachi, N, Ezaki, Y, Liu, J, Watabe, M, Altanshagai, G, Enkhbaatar, B, Dorjnamjaa, D, 'Earliest known Cambrian calcimicrobial reefs occur in the Gobi-Altai, western Mongolia: intriguing geo-biological products immediately after the Ediacaran-Cambrian boundary', *Global and Planetary Change*, 203, 103530 (2021)

Al-Khalili, Jim, McFadden, Johnjoe, *Life on the Edge: The Coming of Age of Quantum Biology*, Crown Publishers (2014)

Cobb, M, *The Egg and Sperm Race*, Free Press (2006)

Darwin, Charles, *On the Origin of Species by Means of Natural Selection, or, The Preservation of Favoured Races in the Struggle for Life*, J. Murray (1859)

Darwin, Charles, *The Descent of Man: And Selection in Relation to Sex*, J. Murray (1871)

Darwin, Charles, *The Expression of the Emotions in Man and Animals*, J. Murray (1872)

Fortey, Richard A, *Life, an Unauthorised Biography: a Natural History of the First Four Thousand Million Years of Life on Earth*, HarperCollins (1997)

Gould, Stephen Jay, *Ontogeny and Phylogeny*, Belknap Press of Harvard University Press (1977)

Haeckel, E, 'Die Entstehung der Sexualzellen bei den Hydromedusen; zugleich ein Beitrag zur Kenntnis des Baues und der Lebenserscheinungen dieser Gruppe', *Archiv für Mikroskopische Anatomie und Entwicklungsmechanik*, 9(2), 263–382 (1873)

Jones, Steve, *Almost Like A Whale: The Origin Of Species Updated*, new edn, Black Swan (2000)

Lane, Nick, *Oxygen: The Molecule That Made The World*, Oxford University Press (2002)

Lane, Nick, *Power, Sex, Suicide: Mitochondria and the Meaning of Life*, Oxford University Press (2005)

Lane, Nick, *Life Ascending: The Ten Great Inventions of Evolution*, Profile Books (2009)

Lane, Nick, *The Vital Question: Why Is Life The Way It Is?*, Profile Books (2015)

Long, John A, *The Dawn of the Deed: The Prehistoric Origins of Sex*, The University of Chicago Press (2012)

Lovejoy, AO, *The Great Chain of Being: A Study of the History of an Idea*, Harvard University Press (1936)

McMenamin, Mark AS, *The Garden of Ediacara: Discovering the First Complex Life*, Columbia University Press (2000)

Moczydłowska, M, 'Taxonomic review of some Ediacaran acritarchs from the Siberian Platform', *Precambrian Research*, 136(3–4), 283–307 (2005)

Plickert, G, Frank, U, Mükker, WA, 'The Hydra Model System', *International Journal of Developmental Biology*, 56, 519–534 (2012)

Rutherford, Adam, *Creation: How Science Is Reinventing Life Itself*, Current (2013)

Sloan, Phillip, *Evolutionary Thought Before Darwin*, The Stanford Encyclopedia of Philosophy (Winter 2019 edition), Edward N. Zalta (ed.) (2019)

Smith, J, Johnson, AB, Thompson, CD, 'Evolutionary effects of mitochondrial segregation and mutation in the development of complex multicellular life with germ line', *Journal of Evolutionary Biology*, 35(3), 412–428 (2022)

Weismann, A, *Die Entstehung der Sexualzellen bei Hydromedusen* [*The origin of the sexual cells in hydromedusae*], Gustav Fischer (1883)

Weismann, A, *Essays upon Heredity*, Clarendon Press (1889)

Woodruff, TK, 'Making eggs: Is it now or later?', *Nature Medicine*, 14(11), 1190–1191 (2008)

3. The early womb

Adonin, L, Drozdov, A, Barlev, NA, 'Sea urchin as a universal model for studies of gene networks', *Frontiers in Genetics*, Volume 11 (20 January 2021)

Anderson, Gemma, 'Endangered: A study of morphological drawing in zoological taxonomy', *Leonardo*, 47(3), 232–213 (2014)

Baltz, JM, Katz, DF, Cone, RA, 'Mechanics of sperm-egg interaction at the zona pellucida', *Biophysical Journal*, 54(4), 643–654 (1988)

Bowler, P, Pickstone, J (eds), *The Cambridge History of Science*, Cambridge University Press (2009)

Carroll, EJ Jr, Acevedo-Duncan, M, Justice, RW, Santiago, L, 'Structure, assembly and function of the surface envelope (fertilization envelope) from eggs of the sea urchin, Strongylocentrotus purpuratus', *Advances in Experimental Medicine and Biology*, 207, 261–291 (1986)

Fortey, Richard A, *Trilobite!: Eyewitness to Evolution*, 1st American edn, Alfred Knopf (1997)

Fox, Douglas, 'What sparked the Cambrian explosion?', *Scientific American*, retrieved from www.scientificamerican.com/article/what-sparked-the-cambrian-explosion (2016)

Geraldes, V, Pinto, E, 'Mycosporine-like amino acids (MAAs): Biology, chemistry and identification features', *Pharmaceuticals (Basel)*, 14(1), 63 (2021)

Gould, Stephen Jay, *Ontogeny and Phylogeny*, Harvard University Press (1977)

Gould, Stephen Jay, *Hen's Teeth and Horse's Toes*, W.W. Norton (1983)

Gould, Stephen Jay, *The Flamingo's Smile*, W.W. Norton (1985)

Gould, Stephen Jay, *Time's Arrow, Time's Cycle*, Harvard University Press (1987)

Gould, Stephen Jay, *Wonderful Life: The Burgess Shale and the Nature of History*, W.W. Norton (1989)

Hörstadius, S, 'The mechanics of sea urchin development: Studies by operative methods', *Biological Reviews*, 14, 132–179 (1939)

Lan, T, Zhao, Y, Esteve, J, Zhao, F, Martínez, P, 'Eggs with trilobite larvae in a Cambrian community. Modeling the hydrodynamics', 10.21203/rs.3.rs-558709/v1 (2021)

Martin, E, 'The egg and the sperm: How science has constructed a romance based on stereotypical male-female roles', *Signs*, 16(3), 485–501 (1991)

Miyake, K, McNeil, PL, 'A little shell to live in: Evidence that the fertilization envelope can prevent mechanically induced damage of the developing sea urchin embryo', *The Biological Bulletin*, 195(2), 214–215 (1998)

Monroy, A, Groeben, C, 'The 'new' embryology at the Zoological Station and at the Marine Biological Laboratory', *Biological Bulletin*, 168, 35–43 (1985)

Pehrson, JR, Cohen, LH, 'The fate of the small micromeres in sea urchin development', *Developmental Biology*, 113(2), 522–526 (1986)

Rojas, J, Hinostroza, F, Vergara, S, Pinto-Borguero, I, Aguilera, F, Fuentes, R, Carvacho, I, 'Knockin' on egg's door: Maternal control of egg activation that influences cortical granule exocytosis in animal species', *Frontiers in Cell and Developmental Biology*, 9 (2021)

Saudemont, A, Haillot, E, Mekpoh, F, Bessodes, N, Quirin, M, Lapraz, F, Duboc, V, Rottinger, E, Range, R, Oisel, A et al., 'Ancestral regulatory circuits governing ectoderm patterning downstream of Nodal and BMP2/4 revealed by gene regulatory network analysis in an echinoderm', *PLoS Genetics*, 6, e1001259 (2010)

Shiba, K, Ohmuro, J, Mogami, Y, Nishigaki, T, Wood, C, Darszon, A, Tatsu, Y, Yumoto, N, Baba, S, 'Sperm-activating peptide induces asymmetric flagellar bending in sea urchin sperm', *Zoological Science*, 22, 293–299 (2005)

Vannier, J, Aria, C, Taylor, RS, Caron, J-B, 'Waptia fieldensis Walcott, a mandibulate arthropod from the middle Cambrian Burgess Shale', *Royal Society Open Science*, 5(6), 172206 (2018)

Zhang, X-g, Pratt, BR, 'Middle Cambrian arthropod embryos with blastomeres', *Science*, 266(5185), 637–639 (1994)

4. Starbursts on shores

The giant petri dish experiment, set up by Harvard Medical School doctors to demonstrate how easily bacteria can evolve antibiotic resistance, can be viewed here: https://www.youtube.com/watch?v=plVk4NVIUh8. I stand by what I said: it's a spectacularly effective way of communicating how natural selection picks away at environmental problems.

Briggs, DEG, Rolfe, WDI, Brannan, J, 'A giant myriapod trail from the Namurian of Arran, Scotland', *Palaeontology*, 22, 273–291 (1979)

Briggs, D, Siveter, DJ, Sutton, MD, Garwood, RJ, Legg, D, 'Silurian horseshoe crab illuminates the evolution of arthropod limbs', *Proceedings of the National Academy of Sciences USA*, 109(39), 15702–15705 (2012)

Fabre, JH, *The Life of the Spider*, Dodd, Mead & Company (1917)

Foster, Russell G, Kreitzman, Leon, *Seasons of Life: The Biological Rhythms That Enable Living Things to Thrive and Survive*, Yale University Press (2009)

Garwood, R et al., 'Early terrestrial animals, evolution and uncertainty', *Evolution: Education and Outreach*, 4(3), 489–501 (2011)

Judson, Olivia, *Dr. Tatiana's Sex Advice to All Creation*, Metropolitan Books (2002)

Lamsdell, J et al., 'The systematics and phylogeny of the Stylonurina (Arthropoda: Chelicerata: Euryptcrida)', *Journal of Systematic Palaeontology*, 8(1), 49–61 (2010)

Lamsdell, J, 'Revised systematics of Palaeozoic horseshoe crabs and the myth of monophyletic Xiphosura', *Zoological Journal of the Linnean Society*, 167(1), 1–27 (2013)

Legg, D, Sutton, MD, Edgecombe, GD, 'Arthropod fossil data increase congruence of morphological and molecular phylogenies', *Nature Communications*, 4(2485) (2013)

Legg, G, 'The external morphology of a new species of ricinuleid (Arachnida) from Sierra Leone', *Journal of Zoology*, 59(1), 1–58 (1976)

Monod, L, Cauwet, L, González-Santillán, E, Huber, S, 'The male sexual apparatus in the order Scorpiones (Arachnida): a comparative study of functional morphology as a tool to define hypotheses of homology', *Frontiers in Zoology*, 14(51) (2017)

Oldfield, G et al., 'Discovery and characterization of spermato-phores in the Eriophyoidea (Acari)', *Annals of the Entomological Society of America*, 63(2), 520–526 (1970)

Peretti, Alfredo, 'An ancient indirect sex model: single and mixed patterns in the evolution of scorpion genitalia', in *The Evolution of Primary Sexual Characters in Animals*, Janet Leonard and Alex Córdoba-Aguilar (eds), Oxford University Press, 218–248 (2010)

Regier, Jerome et al., 'Arthropod relationships revealed by phylo-genomic analysis of nuclear protein-coding sequences', *Nature*, 463, 1079–1083 (2010)

Rosenberg, Gary, 'Independent evolution of terrestriality in Atlantic Truncatellid gastropods', *Evolution*, 50(2), 682–693 (1996)

Schausberger, P, Hoffmann, D, 'Maternal manipulation of hatch-ing asynchrony limits sibling cannibalism in the predatory mite Phytoseiulus persimilis', *Journal of Animal Ecology*, 77, 1109–1114 (2008)

Schultz, J, 'A phylogenetic analysis of the arachnid orders based on morphological characters', *Zoological Journal of the Linnean Society*, 150(2), 221–265 (2007)

Solem, GA, *The Shell Makers: Introducing Molluscs*, John Wiley & Sons, (1974)

Whyte, M, 'A gigantic fossil arthropod trackway', *Nature*, 438, 576 (2005)

Wilson, HM, Anderson, LI, 'Morphology and taxonomy of Paleozoic millipedes (Diplopoda: Chilognatha: Archipolypoda) from Scotland', *Journal of Paleontology*, 78(1), 169–184 (2004)

5. A tale of two fishes

Adams, KR, Fetterplace, LC, Davis, AR, Taylor, MD, Knott, NA, 'Sharks, rays and abortion: the prevalence of capture-induced parturition in elasmobranchs', *Biological Conservation*, 217, 11–27 (2018)

Ishimatsu, A, Yoshida, Y, Itoki, N, Takeda, T, Lee, HJ, Graham, JB, 'Mudskippers brood their eggs in air but submerge them for hatching', *Journal of Experimental Biology*, 210(22), 3946–3954 (2007)

Benton, MJ, *Vertebrate Palaeontology*, 3rd edn, Fig. 7.13 on p. 185, Blackwell (2005)

Buddle, AL, Van Dyke, JU, Thompson, MB, Simpfendorfer, CA, Murphy, CR, Day, ML, Whittington, CM, 'Structure and permeability of the egg capsule of the placental Australian sharp-nose shark, Rhizoprionodon taylori', *Journal of Comparative Physiology B*, 192(2), 263–273 (2022)

Camhi, MD, Valenti, SV, Fordham, SV, Fowler, SL, Gibson, C (eds), *The Conservation Status of Pelagic Sharks and Rays*, Pelagic Shark Red List Workshop, IUCN Shark Specialist Group (2007)

Chapman, DD, Wintner, SP, Abercrombie, DL, Ashe, J, Bernard, AM, Shivji, MS, Feldheim, KA, 'The behavioural and genetic mating system of the sand tiger shark, Carcharias taurus, an intrauterine cannibal', *Biology Letters*, 9(3), 20130003 (2013)

Fortey, Richard A, *Life, an Unauthorised Biography: a Natural History of the First Four Thousand Million Years of Life on Earth*, HarperCollins (1998)

Gess, RW, Coates, MI, 'Fossil coelacanths from the South African Devonian', *Zoological Journal of the Linnean Society*, 175, 360–383 (2015)

Hamlett, E, 'Uterogestation and placentation in elasmobranchs', *Journal of Experimental Zoology*, 266, 347–369 (1993)

Hiramatsu, N, Todo, T, Sullivan, CV, Schilling, J, Reading, BJ, Matsubara, T, Ryu, YW, Mizuta, H, Luo, W, Nishimiya, O, Wu, M, Mushirobira, Y, Yilmaz, O, Hara, A, 'Ovarian yolk formation in fishes: molecular mechanisms underlying formation of lipid droplets and vitellogenin-derived yolk proteins', *General Comparative Endocrinology*, 15(221), 9–15 (2015)

Hosken, DJ, House, CM, 'Sexual selection', *Current Biology*, 21(2), R62–R65.

Iida, A, Nishimaki, T, Sehara-Fujisawa, A, 'Prenatal regression of the trophotaenial placenta in a viviparous fish, Xenotoca eiseni', *Scientific Reports*, 5, 7855 (2015)

Ishimatsu, A, Hishida, Y, Takita, T, Kanda, T, Oikawa, S, Takeda, T, Huat, K, 'Mudskippers store air in their burrows', *Nature*, 391, 237–238 (1998)

Joachimski, MM, Breisig, S, Buggisch, W, Talent, JA, Mawson, R, Gereke, M, Morrow, JR, Day, J, Weddige, K, 'Devonian climate and reef evolution: insights from oxygen isotopes in apatite', *Earth and Planetary Science Letters*, 284(3–4), 599–609 (2009)

Sallan, LC, Coates, MI, 'The long-rostrumed elasmobranch Bandringa Zangerl, 1969, and taphonomy within a Carboniferous shark nursery', *Journal of Vertebrate Paleontology*, 34(1), 22–33 (2014)

Lennon, E, Philips, N, Garbett, A, Carlsson, J, Carlsson, J, Crowley, D, Judge, M, Yeo, I, Collins, P, 'Going deeper, darker and further: observations charting an egg nursery, a range and depth extension for the deep-sea spiny tailed skate *Bathyraja spinicauda*, first records from the Mid Atlantic Ridge', *Deep Sea Research Part I, Oceanographic Research Papers*, 175, Article 103584 (2021)

Lindsay, WR, Andersson, S, Bererhi, B, Höglund, J, Johnsen, A, Kvarnemo, C, Leder, EH, Lifjeld, JT, Ninnes, CE, Olsson, M, Parker, GA, Pizzari, T, Qvarnström, A, Safran, RJ, Svensson, O, Edwards, SV, 'Endless forms of sexual selection', *PeerJ*, 7, e7988 (2019)

Long, John, *Swimming in Stone: the Amazing Gogo Fossils of the Kimberley*, Fremantle Arts Centre (2007)

Long, John A, *The Dawn of the Deed: The Prehistoric Origins of Sex*, The University of Chicago Press (2012)

McGrath, C, 'Highlight: big surprises from the world's smallest fish', *Genome Biology and Evolution*, 10(4), 1104–1105 (2018)

Miya, M, Nemoto, T, 'Reproduction, growth and vertical distribution of the meso- and bathypelagic fish Cyclothone atraria (Pisces: Gonostomatidae) in Sagami Bay, Central Japan', *Deep Sea Research Part A, Oceanographic Research Papers*, 34(9), 1565–1577 (1987)

Naylor, ER, Kawano, SM, 'Mudskippers modulate their locomotor kinematics when moving on deformable and inclined substrates', *Integrative and Comparative Biology*, 9, icac084 (2022)

Nelson, JS, Grande, TC, Wilson, MVH, 'Classification of fishes', in *Fishes of the World*, 5th edn, Wiley (2016)

Niedźwiedzki, G, Szrek, P, Narkiewicz, K, Narkiewicz, M, Ahlberg, P, 'Tetrapod trackways from the Middle Devonian Period of Poland', *Nature*, 463, 43–48 (2010)

Otani, S, Iwai, T, Nakahata, S, Sakai, C, Yamashita, M, 'Artificial fertilization by intracytoplasmic sperm injection in a teleost fish, the medaka (Oryzias latipes)', *Biology of Reproduction*, 80, 175–183 (2008)

Polgar, G, Fuchino, R, 'English translation of Kobayashi et al. (1972) Egg development and rearing experiments of the larvae of the mud skipper *Periophthalmus cantonensis*', *Bulletin of the Faculty of Fisheries*, Nagasaki University, 33, 49–62 (2014)

Shubin, N, *Your Inner Fish: a Journey into the 3.5-Billion-Year History of the Human Body*, Vintage Books (2009)

Wicaksono, A, Hidayat, S, Damayanti, Y, Jin, DSM, Sintya, E, Retnoaji, B, Alam, P, 'The significance of pelvic fin flexibility for tree climbing fish', *Zoology*, 119(6), 511–517 (2016)

Wourms, JP, 'Reproduction and development in chondrichthyan fishes', *American Zoologist*, 17, 379–410 (1977)

Wourms, JP, 'Viviparity: the maternal-fetal relationship in fishes', *American Zoologist*, 21, 473–515 (1981)

Wourms, JP, Grove, BD, Lombardi, J, 'The maternal-embryonic relationship in viviparous fishes', in *Fish Physiology*, vol. XI, WS Hoar, DJ Randall (eds), Academic Press, 1–134 (1988)

van der Laan, R, Eschmeyer, WN, Fricke, R, 'Family-group names of recent fishes', *Zootaxa*, 3882(2), 1–230 (2014)

Zúñiga-Vega, JJ, Aspbury, AS, Johnson, JB, Pollux, BJA, 'Editorial: ecology, evolution, and behavior of viviparous fishes', *Frontiers in Ecology and Evolution*, 10 (2022)

6. A most marvellous invention

Bayer, C, Zhou, X, Zhou, B, Riddiford, LM, von Kalm, L, 'Evolution of the Drosophila broad locus: the Manduca sexta broad Z4 isoform has biological activity in Drosophila', *Development Genes and Evolution*, 213(10), 471–476 (2003)

Benton, MJ, *Vertebrate Palaeontology*, 4th edn, Wiley-Blackwell (2014)

Carroll, RL, *Vertebrate Paleontology and Evolution*, W.H. Freeman (1988)

Clack, JA, *Gaining Ground: The Origin and Evolution of Tetrapods*, Indiana University Press (2002)

Fabre, JH, *The Life of the Fly*, Dodd, Mead & Company (1921)

Fortey, Richard A. *Life, an Unauthorised Biography: a Natural History of the First Four Thousand Million Years of Life on Earth*, HarperCollins (1998)

Gould, Stephen Jay, *Ever Since Darwin*, W.W. Norton (1977)

Gould, Stephen Jay, *The Panda's Thumb*, W.W. Norton (1980)

Haeckel, E, *Generelle Morphologie der Organismem*, 2 vols, G. Reimer (1866)

Holmes, Tao Tao, *The Blobby, Dazzling World of Insect Eggs*, Atlas Obscura (2016)

Reisz, RR, 'The origin and early evolutionary history of amniotes', *Trends in Ecology and Evolution*, 12, 218–222 (1997)

Shear, W, Kukalová-Peck, J, 'The ecology of Paleozoic terrestrial arthropods: the fossil evidence', *Canadian Journal of Zoology-revue Canadienne De Zoologie*, 68, 1807–1834 (1990)

Sues, HD, *The Rise of Reptiles: 320 Million Years of Evolution*, Johns Hopkins University Press (2019)

Sumida, SS, Martin, KLM, *Amniote Origins: Completing the Transition to Land*, Academic Press (1997)

7. The larval storm

Piñeiro, G, Ferigolo, J, Meneghel, M, Laurin, M, 'The oldest known amniotic embryos suggest viviparity in mesosaurs', *Historical Biology*, 24(6), 620–630 (2012)

Hardy, ICW, 'A macroevolutionary fondness for Neoptera', *Trends in Ecology and Evolution*, 17(8), 354 (2002)

Hitt, Jack, 'The battle over the sea-monkey fortune', *The New York Times*, archived from the original on December 25, 2018, retrieved 16 April 2016

Hoddle, MS, Van Driesche, RG, Sanderson, JP, 'Biology and use of the whitefly parasitoid Encarsia formosa', *Annual Review of Entomology*, 43(1), 645–669 (1998)

Huber, JT, Beardsley, JW, 'A new genus of fairyfly, Kikiki, from the Hawaiian Islands (Hymenoptera: Mymaridae)', *Proceedings of the Hawaiian Entomological Society*, 34, 65–70 (2000)

Huber, JT, Noyes, JS, 'A new genus and species of fairyfly, Tinkerbella nana (Hymenoptera, Mymaridae), with comments on its sister genus Kikiki, and discussion on small size limits in arthropods', *Journal of Hymenoptera Research*, 32, 17–44 (2013)

Katona, G, Szabó, F, Végvári, Z, Székely, T Jr, Liker, A, Freckleton, RP, Vági, B, Székely, T, 'Evolution of reproductive modes in sharks and rays', *Journal of Evolutionary Biology*, 36(11), 1630–1640 (2023)

McAlister, Erica, *The Secret Life of Flies*, CSIRO Publishing (2017)

Mitchell, FL, Lasswell, J, *A Dazzle of Dragonflies*, Texas A&M University Press, p. 47 (2005)

Penney, D, Jepson, JE, *Fossil Insects: An Introduction to Palaeoentomology*, Siri Scientific Press, p. 79 (2014)

Peters, RS, Meusemann, K, Petersen, M et al., 'The evolutionary history of holometabolous insects inferred from transcriptome-based phylogeny and comprehensive morphological data', *BMC Evolutionary Biology*, 14(52) (2014)

Peters, R et al., 'Evolutionary history of the Hymenoptera', *Current Biology*, 27, 1013–1018 (2017)

Retallack, GJ, Krull, ES, 'Carbon isotopic evidence for terminal-Permian methane outbursts and their role in extinctions of animals, plants, coral reefs, and peat swamps', *Geological Society of America Special Paper*, 399, 249 (2006)

Summer, S, *Endless Forms: The Secret World of Wasps*, HarperCollins (2022)

Sverdrup-Thygeson, Anne, Moffatt, Lucy. *Extraordinary Insects: Weird, Wonderful, Indispensable, the Ones Who Run Our World*, Trans. Lucy Moffatt, Mudlark (2019)

Tajik, H, Moradi, M, Rohani, SM, Erfani, AM, Jalali, FS, 'Preparation of chitosan from brine shrimp (Artemia urmiana) cyst shells and effects of different chemical processing sequences on the physicochemical and functional properties of the product', *Molecules*, 13(6), 1263–1274 (2008)

Truman, JW, Riddiford, LM, 'Endocrine insights into the evolution of metamorphosis in insects', *Annual Review of Entomology*, 47, 467–500 (2002)

8. The Triassic takeover

Avanzini, M, Vecchia, FMD, Mietto, P, Piubelli, D, Preto, N, Rigo, M, Roghi, G, 'A vertebrate nesting site in Northeastern Italy reveals unexpectedly complex behavior for late Carnian reptiles', *PALAIOS*, 22(5), 465–475 (2007)

Bücker, H, Horneck, G, 'The biological effectiveness of HZE-particles of cosmic radiation studied in the Apollo 16 and 17 Biostack experiments', *Acta Astronautica*, 2(3–4), 247–264 (1975)

Carroll, Robert L, *The Rise of Amphibians: 365 Million Years of Evolution*, Johns Hopkins University Press (2009)

Collin, Rachel, Leonard, Janet L (eds), 'Transitions in sexual and reproductive strategies among the Caenogastropoda', in *Transitions Between Sexual Systems: Understanding the Mechanisms of, and Pathways Between, Dioecy, Hermaphroditism and Other Sexual Systems*, Springer International Publishing, 193–220 (2018)

Frost, Darrel R, Grant, Taran, Faivovich, Julián, Bain, Raoul H, Haas, Alexander, Haddad, Célio FB, De Sá, Rafael O, Channing, Alan, Wilkinson, Mark, Donnellan, Stephen C, Raxworthy, Christopher J, Campbell, Jonathan A, Blotto, Boris L, Moler, Paul, Drewes, Robert C, Nussbaum, Ronald A, Lynch, John D, Green, David M, Wheeler, Ward C, 'The amphibian tree of life', *Bulletin of the American Museum of Natural History*, 297, 1–291 (2006)

Hayward, JL, Hirsch, KF, Robertson, TC, 'Rapid dissolution of avian eggshells buried by Mount St. Helens ash', *PALAIOS*, 6, 174–178 (1991)

Hirsch, KF, 'The fossil record of vertebrate eggs', in *The Palaeobiology of Trace Fossils*, SK Donovan (ed.), John Wiley and Sons, 269–294 (1994)

Jacobs, Louis L, 'African dinosaurs', in *Encyclopedia of Dinosaurs*, Phillip J Currie, Kevin Padian (eds), Academic Press, 2–4 (1997)

Mainwaring, Mark C, Stoddard, Mary, Caswell, Barber Iain, Deeming, D Charles, Hauber, Mark E, 'The evolutionary ecology of nests: a cross-taxon approach', *Philosophical Transactions of the Royal Society B*, 3782022013620220136 (2023)

Marzoli, A et al., 'Extensive 200-million-year-old continental flood basalts of the Central Atlantic Magmatic Province', *Science*, 284, 618–620 (1999)

Mazaheri-Johari, Mina, Roghi, Guido, Caggiati, Marcello, Kustatscher, Evelyn, Ghasemi-Nejad, Ebrahim, Zanchi, Andrea, Gianolla, Piero, 'Disentangling climate signal from tectonic forcing: The Triassic Aghdarband Basin (Turan Domain, Iran)', *Palaeogeography, Palaeoclimatology, Palaeoecology*, 586, 110777 (2022)

Norell, MA, Wiemann, J, Fabbri, M et al., 'The first dinosaur egg was soft', *Nature*, 583, 406–410 (2020)

Stubbs, TL, Pierce, SE, Elsler, A, Anderson, PSL, Rayfield, EJ, Benton, MJ, 'Ecological opportunity and the rise and fall of crocodylomorph evolutionary innovation', *Proceedings of the Royal Society of London Series B*, 288, 1947 (2021)

Tokaryk, TT, Storer, J, 'Dinosaur eggshell fragments from Saskatchewan, and evaluation of potential distance of eggshell transport', *Journal of Vertebrate Paleontology*, 11(3), 58A (1991)

Vermeij, Geerat J, Dudley, Robert, 'Why are there so few evolutionary transitions between aquatic and terrestrial ecosystems?', *Biological Journal of the Linnean Society*, 70(4), 541–554 (2000)

Wilberg, Eric W, Turner, Alan H, Brochu, Christopher A, 'Evolutionary structure and timing of major habitat shifts in Crocodylomorpha', *Scientific Reports*, 9(1), 514 (2019)

9. Navel-gazing

Archer, M, Jenkins, FA Jr, Hand, SJ, Murray, P, Godthelp, H, 'Description of the skull and non-vestigial dentition of a Miocene platypus (Obdurodon dicksoni n. sp.) from Riversleigh, Australia, and the problem of monotreme origins', in *Platypus and Echidnas*, ML Augee (ed.), Royal Zoological Society of New South Wales, 15–27 (1992)

Archer, M, Flannery, TM, Ritchie, A, Molnar, RE, 'First Mesozoic mammal from Australia – an early Cretaceous monotreme', *Nature*, 318, 363–366 (1985)

Crichton, AI, Worthy, TH, Camens, AB, Prideaux, GJ, 'A new ektopodontid possum (Diprotodontia, Ektopodontidae) from the Oligocene of central Australia, and its implications for phalangeroid interrelationships', *Journal of Vertebrate Paleontology*, 42(3) (2022)

Ashby, J, *Platypus Matters: The Extraordinary Story of Australia's Mammals*, HarperCollins (2022)

Brusatte, Stephen, *The Rise and Reign of the Mammals: A New History, from the Shadow of the Dinosaurs to Us*, Mariner Books (2022)

Brusatte, SL, Benton, MJ, Ruta, M, Lloyd, GT, 'Superiority, competition, and opportunism in the evolutionary radiation of dinosaurs', *Science*, 321(5895), 1485–1488 (2008)

Carroll, Robert L, *Vertebrate Paleontology and Evolution*, W.H. Freeman (1988)

Celik, Mélina, Cascini, Manuela, Haouchar, Dalal, Van Der Burg, Chloe, Dodt, William, Evans, Alistair, Prentis, Peter, Bunce, Michael, Fruciano, Carmelo, Phillips, Matthew, 'A molecular and morphometric assessment of the systematics of the Macropus complex clarifies the tempo and mode of kangaroo evolution', *Zoological Journal of the Linnean Society*, 186(3), 793–812 (2019)

Croft, DB, *Marsupialia (Marsupials)*, John Wiley & Sons (2021)

Dawson TJ, *Kangaroos: Biology of the Largest Marsupials*, Cornell University Press (1995)

Dawson, L, Flannery, T, 'Taxonomic and phylogenetic status of living and fossil kangaroos and wallabies of the genus Macropus Shaw (Macropodidae: Marsupialia), with a new subgeneric name for the larger wallabies', *Australian Journal of Zoology*, 33(4), 473–498 (1985)

Flannery, TF, *Chasing Kangaroos: A Continent, a Scientist, and a Search for the World's Most Extraordinary Creature*, 1st American edn, Grove (2008)

Flannery TF, *The Future Eaters: An Ecological History of the Australasian Lands and People*, Grove Press, 67–75 (2002)

Flannery, TF, Archer, M, Rich, TH, Jones, R, 'A new family of monotremes from the Cretaceous of Australia', *Nature*, 377, 418–420 (1995)

Fortey, Richard A, *Life, an Unauthorised Biography: a Natural History of the First Four Thousand Million Years of Life on Earth*, HarperCollins (1998)

Fortey, Richard A, *Dry Store Room No. 1: The Secret Life of the Natural History Museum*, HarperPress (2008)

Gould, Stephen Jay, *The Lying Stones of Marrakech*, Harmony Books (2000)

Grellet-Tinner, Gerald, Wroe, Stephen, Thompson, Michael, Ji, Qiang, 'A note on pterosaur nesting behavior', *Historical Biology*, 19, 273–277 (2007)

Griffiths, M, *The Biology of the Monotremes*, Academic Press (1978)

Griffiths, M, Wells, RT, Barrie, DJ, 'Observations on the skulls of fossil and extant echidnas (Monotremata: Tachyglossidae)', *Australian Mammalogy*, 14, 87–101 (1991)

Jackson, S, Groves, C, *Taxonomy of Australian Mammals*, CSIRO Publishing, p. 157 (2015)

Johnson, MH, Everitt, BJ, *Essential Reproduction*, Blackwell Scientific (1988)

Jones, M, Dickman, C, Archer, M, *Predators with Pouches: The Biology of Carnivorous Marsupials*, Collingwood (2003)

Knobil, E, Neill, JD (eds), *Physiology of Reproduction*, vol. 3, Academic Press (1998)

Linville, BJ, Stewart, JR, Ecay, TW, Herbert, JF, Parker, SL, Thompson, MB, 'Placental calcium provision in a lizard with prolonged oviductal egg retention', *Journal of Comparative Physiology – Biochemical Systemic and Environmental Physiology*, 180, 221–227 (2010)

Long, JA et al., *Prehistoric Mammals of Australia and New Guinea: One Hundred Million Years of Evolution*, Johns Hopkins University Press (2002)

Marshall, M, 'Live birth, evolving before our eyes', *New Scientist*, 25 August 2010

McCullough, DR, McCullough, Y, *Kangaroos in Outback Australia: Comparative Ecology and Behavior of Three Coexisting Species*, Columbia University Press (2000)

Pascual, R, Archer, M, Juareguizar, EO, Prado, JL, Godthelp, H, Hand, SJ, 'First discovery of monotremes in South America', *Nature*, 356, 704–706 (1992)

Reisz, RR, Evans, DC, Roberts, EM, Sues, HD, Yates, AM, 'Oldest known dinosaurian nesting site and reproductive biology of the Early Jurassic sauropodomorph Massospondylus', *Proceedings of the National Academy of Sciences USA*, 109(7), 2428–2433 (2012)

Qualls, C, Andrews, R, 'Cold climates and the evolution of viviparity in reptiles: cold incubation temperatures produce poor-quality offspring in the lizard, Sceloporus virgatus', *Biological Journal of the Linnean Society*, 67(3), 353–376 (1999)

Roberts, RM, Green, JA, Schulz, LC, 'The evolution of the placenta', *Reproduction (Cambridge, England)*, 152(5), R179–R189 (2016)

Shaw, George, Nodder, Frederick Polydore, 'The duck-billed platypus, Platypus anatinus', *The Naturalist's Miscellany*, 10(CXVIII), 385–386 (1799)

Smith, KK, Keyte, AL, 'Adaptations of the marsupial newborn: birth as an extreme environment', *Anatomical Record*, 303, 235–249 (2020)

Taylor, AC, Taylor, P, 'Sex of pouch young related to maternal weight in Macropus eugenii and M. parma', *Australian Journal of Zoology*, 45(6), 573–578 (1997)

Tyndale-Biscoe, CH, 'Australasian marsupials–to cherish and to hold', *Reproduction, Fertility and Development*, 13(7–8), 477–485 (2001)

Van Dyke, JU, Brandley, MC, Thompson, MB, 'The evolution of viviparity: molecular and genomic data from squamate reptiles advance understanding of live birth in amniotes', *Reproduction*, 147(1), R15–R26 (2014)

Wellnhofer, Peter, *The Illustrated Encyclopedia of Pterosaurs: An Illustrated Natural History of the Flying Reptiles of the Mesozoic Era*, Crescent Books (1991)

Witton, Mark, *Pterosaurs: Natural History, Evolution, Anatomy*, Princeton University Press (2013)

10. The art in the isthmus

This chapter of the book, especially, owes an extraordinary debt to Tim Birkhead, author of *The Most Perfect Thing*, whose engaging science writing is the best out there and whose technical knowledge of the birds' egg is, in my eyes, unmatched. If you are now inspired by oology, his book (which is primarily about bird eggs) is the next you should read. I would like to thank Tim personally – this chapter could not have been written without him.

Some readers may like to see a visual representation of what it is like to journey through a bird's oviduct and so, if this is you, I am happy to oblige. A few years ago, I was lucky enough to work with biologist Patricia Brennan and produce a virtual reality 'tour' of the complicated reproductive anatomy of a bird, the Muscovy duck. The experience, called *Duck Genitalia Explorer*, is free to download on Google Play via a mobile phone.

Amiot, R, Golovneva, LB, Godefroit, P, Goedert, J, Garcia, G, Lécuyer, C, Fourel, F, Herman, AB, Spicer, AB, 'High-latitude dinosaur nesting strategies during the Latest Cretaceous in North-Eastern Russia', *Diversity*, 15, 565 (2023)

Bailleul, AM, O'Connor, J, Zhang, S et al., 'An Early Cretaceous enantiornithine (Aves) preserving an unlaid egg and probable medullary bone', *Nature Communications*, 10, 1275 (2019)

Birkhead, Tim, *The Most Perfect Thing: Inside (and Outside) a Bird's Egg*, Bloomsbury USA (2016)

Birkhead, T, Van Grouw, D, *Bird Sense: What It's Like to Be a Bird*, Bloomsbury (2012)

Black, Riley, *My Beloved Brontosaurus: On the Road with Old Bones, New Science, and Our Favorite Dinosaurs*, Scientific American/ Farrar, Straus and Giroux (2013)

Black, Riley, *Written in Stone: Evolution, the Fossil Record, and Our Place in Nature*, Bellevue Literary Press (2010)

Deeming, Charles D, Mayr, Gerald, 'Pelvis morphology suggests that early Mesozoic birds were too heavy to contact incubate their eggs', *Journal of Evolutionary Biology*, 31, 701–709 (2018)

Choh, Y, Janssen, A, 'A tiny cuckoo: risk-dependent interspecific brood parasitism in a predatory mite', *Functional Ecology*, 37, 1594–1603 (2023)

Duda, MP, Grooms, C, Sympson, L, Blais, JM, Dagodzo, D, Feng, W, Hayward, KM, Julius, ML, Kimpe, LE, Lambertucci, SA,

Layton-Matthews, D, Lougheed, SC, Massaferro, J, Michelutti, N, Pufahl, PK, Vuletich, A, Smol, JP, 'A 2200-year record of Andean Condor diet and nest site usage reflects natural and anthropogenic stressors', *Proceedings of the Royal Society of London Series B*, 290, 1998 (2023)

Dyke, G, Vremir, M, Kaiser, G, Naish, D, 'A drowned Mesozoic bird breeding colony from the Late Cretaceous of Transylvania', *Naturwissenschaften*, 99, 435–442 (2012)

Fernández, MS, García, RA, Fiorelli, L, Scolaro, A, Salvador, RB, Cotaro, CN, Kaiser, GW, Dyke, GJ, 'A large accumulation of avian eggs from the late Cretaceous of patagonia (Argentina) reveals a novel nesting strategy in mesozoic birds', *PLoS One*, 8(4), e61030 (2013)

Gould, Stephen Jay, *Bully for Brontosaurus*, W.W. Norton (1991)

Gould, Stephen Jay, *Eight Little Piggies*, W.W. Norton (1993)

Gould, Stephen Jay, *Dinosaur in a Haystack*, Harmony Books (1995)

Graveland, J, Berends, AE, 'Timing of the calcium intake and effect of calcium deficiency on behaviour and egg laying in captive great tits, Parus major', *Physiological Zoology*, 70(1), 74–84 (1997)

Hirsch, KF, Kihm, AJ, Zelenitsky, DK, 'New eggshell of ratite morphotype with predation marks from the Eocene of Colorado', *Journal of Vertebrate Paleontology*, 17(2), 360–369 (1997)

Knoll, F, Chiappe, LM, Sanchez, S, Garwood, RJ, Edwards, NP, Wogelius, RA, Sellers, WI, Manning, PL, Ortega, F, Serrano, FJ, Marugán-Lobón, J, Cuesta, E, Escaso, F, Sanz, JL, 'A diminutive perinate European Enantiornithes reveals an asynchronous ossification pattern in early birds', *Nature Communications*, 9(1), 937 (2018)

Liu, J, Yang, C, Yu, J, Wang, H, Pape Møller, A, Liang, W, 'Egg recognition and brain size in a cuckoo host', *Behavioural Processes*, 180, 104223 (2020)

Naish, D, 'Fossil record of bird behaviour', *Journal of Zoology*, 292, 268–280 (2014)

Rasnitsyn, AP, Quicke, DLJ, *History of Insects*, Kluwer Academic Publishers (2002)

Skinner, Brian J, Porter, Stephen C, *The Dynamic Earth: An Introduction to Physical Geology*, 3rd edn, John Wiley & Sons (1995)

Stoddard, MC, Yong, EH, Akkaynak, D, Sheard, C, Tobias, JA, Mahadevan, L, 'Avian egg shape: form, function, and evolution', *Science*, 356(6344), 1249–1254 (2017)

St. Louis, VL, Breebaart, L, 'Calcium supplements in the diet of nestling tree swallows near acid sensitive lakes', *The Condor*, 93(2), 286–294 (1991)

Wiemann, J, Yang, TR, Norell, MA, 'Dinosaur egg colour had a single evolutionary origin', *Nature*, 563, 555–558 (2018)

11. The invasive placenta

Adolfi, MC, Nakajima, RT, Nóbrega, RH, Schartl M, 'Intersex, hermaphroditism, and gonadal plasticity in vertebrates: evolution of the Müllerian duct and signaling', *Annual Review of Animal Bioscience*, 7, 149–151 (2018)

Allen, WR, 'Ovulation, pregnancy, placentation and husbandry in the African elephant (Loxodonta africana)', *Philosophical Transactions of the Royal Society of London. Series B, Biological Sciences*, 361(1469), 821–834 (2006)

Andersen, Nils, Cheng, Lanna, 'The marine insect Halobates (Heteroptera: Gerridae): biology, adaptations, distribution, and phylogeny', *Oceanography and Marine Biology: An Annual Review*, 42, 119–180 (2004)

Bates, JM, Fidino, M, Nowak-Boyd, L, Strausberger, BM, Schmidt, KA, Whelan, CJ, 'Climate change affects bird nesting phenology: Comparing contemporary field and historical museum nesting records', *Journal of Animal Ecology*, 92, 263–272 (2023)

Bennett, Carys E, Thomas, Richard, Williams, Mark, Zalasiewicz, Jan, Edgeworth, Matt, Miller, Holly, Coles, Ben, Foster, Alison, Burton, Emily J, Marume, Upenyu, 'The broiler chicken as a signal of a human reconfigured biosphere', *Royal Society Open Science*, 5, 180325 (2018)

Benton, MJ, *The Rise of the Mammals*, Eagle Editions (1991)

Bertrand, OC, Shelley, SL, Williamson, TE, Wible, JR, Chester, SGB, Flynn, JJ, Holbrook, LT, Lyson, TR, Meng, J, Miller, IM, Püschel, HP, Smith, T, Spaulding, M, Tseng, ZJ, Brusatte, SL, 'Brawn before brains in placental mammals after the end-Cretaceous extinction', *Science*, 376(6588), 80–85 (2022)

Birkhead, Tim, *The Most Perfect Thing: Inside (and Outside) a Bird's Egg*, Bloomsbury USA (2016)

Bohannon, Cat, *Eve: How the Female Body Drove 200 Million Years of Human Evolution*, Knopf Doubleday Publishing Group (2023)

Bradshaw, John, *The Animals Among Us: The New Science of Anthrozoology*, Penguin Books (2017)

Burton, GJ, Fowden, AL, 'The placenta: a multifaceted, transient organ', *Philosophical Transactions of the Royal Society B*, 370(1663), 20140066 (2015)

Carter, AM, Miglino, MA, Ambrosio, CE, Santos, TC, Rosas, FC, Neto, JA, Lazzarini, SM, Carvalho, AF, da Silva, VM, 'Placentation in the Amazonian manatee (Trichechus inunguis)', *Reproduction, Fertility and Development*, 20(4), 537–545 (2008)

Crick, H, Dudley, C, Glue, D et al., 'UK birds are laying eggs earlier', *Nature*, 388, 526 (1997)

Cobb, Matthew, *The Idea of the Brain: A History*, Profile Books (2020)

Darwin, Charles, *The Descent of Man: And Selection in Relation to Sex*, J. Murray (1871)

Dawkins, R, *The Selfish Gene*, Oxford University Press (1989)

Donovan, M, Iglesias, A, Wilf, P et al., 'Rapid recovery of Patagonian plant–insect associations after the end-Cretaceous extinction', *Nature Ecology and Evolution*, 1, 0012 (2017)

Drew, Liam, *I, Mammal: the Story of What Makes Us Mammals*, Bloomsbury Sigma (2019)

Elliot, MG, Crespi, BJ, 'Placental invasiveness and brain-body allometry in eutherian mammals', *Journal of Evolutionary Biology*, 21(6), 1763–1778 (2008)

Ellis, EC, Klein Goldewijk, K, Siebert, S, Lightman, D, Ramankutty, N, 'Anthropogenic transformation of the biomes, 1700 to 2000', *Global Ecology and Biogeography*, 19(5), 589–606 (2010)

Field, Daniel J et al., 'Early evolution of modern birds structured by global forest collapse at the end-Cretaceous mass extinction', *Current Biology*, 28(11), 1825–1831 (2018)

Gaeth, AP, Short, RV, Renfree, MB, 'The developing renal, reproductive, and respiratory systems of the African elephant suggest an aquatic ancestry', *Proceedings of the National Academy of Sciences USA*, 96(10), 5555–5558 (1999)

Garratt, Michael, Gaillard, Jean-Michel, Brooks, Robert C, Lemaitre, JF, 'Diversification of the eutherian placenta is associated with changes in the pace of life', *Proceedings of the National Academy of Sciences USA*, 110, 7760–7765 (2013)

Geist, HJ, Lambin, EF, 'Proximate causes and underlying driving forces of tropical deforestation', *BioScience*, 52(2), 143–150 (2002)

Gibbs, HK, Ruesch, AS, Achard, F, Clayton, MK, Holmgren, P, Ramankutty, N, Foley, JA, 'Tropical forests were the primary sources of new agricultural land in the 1980s and 1990s', *Proceedings of the National Academy of Sciences USA*, 107(38), 16732–16737 (2010)

Grier, JW, 'Ban of DDT and subsequent recovery of reproduction in bald eagles', *Science*, 218, 1232–1234 (1982)

Hosonuma, N, Herold, M, De Sy, V, De Fries, RS, Brockhaus, M, Verchot, L et al., 'An assessment of deforestation and forest degradation drivers in developing countries', *Environmental Research Letters*, 7(4), 044009 (2012)

Järvinen, A, 'Correlation between egg size and clutch size in the Pied Flycatcher *Ficedula hypoleuca* in cold and warm summers', *Ibis*, 138, 620–623 (1996)

Ausubel, JH, Wernick, IK, Waggoner, PE, 'Peak farmland and the prospect for land sparing', *Population and Development Review*, 38, suppl. s1, 221–242 (2013)

Lieberman, Daniel E, *The Story of the Human Body: Evolution, Health, and Disease*, Pantheon Books (2013)

Turco, MY, Moffett, A, 'Development of the human placenta', *Development*, 146 (22): dev163428 (2019)

Marini, Miguel, Hall, Linnea, Bates, John, Steinheimer, Frank, McGowan, Robert, Silveira, Luís, Lijtmaer, Dario, Tubaro, Pablo, Córdoba-Córdoba, Sergio, Gamauf, Anita, Greeney, Harold, Schweizer, Manuel, Kamminga, Pepijn, Cibois, Alice, Vallotton, Laurent, Russell, Douglas, Robinson, Scott, Sweet, Paul, Frahnert, Sylke, Heming, Neander, 'The five million bird eggs in the world's museum collections are an invaluable and underused resource', *The Auk*, 137, 4 (2020)

Maurer, G, Portugal, SJ, Hauber, ME, Mikšík, I, Russell, DGD, Cassey, P, 'First light for avian embryos: eggshell thickness and pigmentation mediate variation in development and UV exposure in wild bird eggs', *Functional Ecology*, 29, 209–218 (2015)

Murphy, WJ, Eizirik, E, O'Brien, SJ, Madsen, O, Scally, M, Douady, CJ et al., 'Resolution of the early placental mammal radiation using Bayesian phylogenetics', *Science*, 294(5550), 2348–2351 (2001)

Poore, J, Nemecek, T, 'Reducing food's environmental impacts through producers and consumers', *Science*, 360(6392), 987–992 (2018)

Pounds, J Alan, Bustamante, Martín R, Coloma, Luis A, Consuegra, Jamie A, Fogden, Michael PL, Foster, Pru N, La Marca, Enrique, Masters, Karen L, Merino-Viteri, Andrés, Puschendorf, Robert, Ron, Santiago R, Sánchez-Azofeifa, G Arturo, Still, Christopher J, Young, Bruce E, 'Widespread amphibian extinctions from epidemic disease driven by global warming', *Nature*, 439(7073), 161–167 (2006)

Sagan, Carl, *Broca's Brain: Reflections on the Romance of Science*, Random House (1979)

Springer, MS, Stanhope, MJ, Madsen, O, de Jong WW, 'Molecules consolidate the placental mammal tree', *Trends in Ecology and Evolution*, 19(8), 430–438 (2004)

Stuart, SN, Hoffmann, M, Chanson, JS, Cox, NA, Berridge, RJ, Ramani, P, Young, BE (eds), *Threatened Amphibians of the World*, Lynx Edicions, in association with IUCN – The World Conservation Union, Conservation International and NatureServe (2008)

Stuart, SN, Chanson, JS, Cox, NA, Young, BE, Rodrigues, ASL, Fischman, DL, Waller, RW, 'Status and trends of amphibian declines and extinctions worldwide', *Science*, 306(5702), 1783–1786 (2004)

Visser, ME, van Noordwijk, AJ, Tinbergen, JM, Lessells, CM, 'Warmer springs lead to mistimed reproduction in great tits (Parus major)', *Proceedings of the Royal Society of London Series B*, 265, 1867–1870 (1998)

Wildman, DE, Chen, C, Erez, O, Grossman, LI, Goodman, M, Romero, R, 'Evolution of the mammalian placenta revealed by phylogenetic analysis', *Proceedings of the National Academy of Sciences USA*, 103(9), 3203–3208 (2006)

Wilson, Edward O, *The Diversity of Life*, new edn, W.W. Norton (1999)

Zalasiewicz, JA, Freedman, K, *The Earth after Us What Legacy Will Humans Leave in the Rocks?*, Oxford University Press (2008)

INDEX